SYSTEMS OF FREQUENCY CURVES

SYSTEMS OF FREQUENCY CURVES

WILLIAM PALIN ELDERTON

AND

NORMAN LLOYD JOHNSON

Professor of Statistics, University of North Carolina, Chapel Hill

CAMBRIDGE
AT THE UNIVERSITY PRESS
1969

CAMBRIDGE UNIVERSITY PRESS
Cambridge, New York, Melbourne, Madrid, Cape Town, Singapore, São Paulo, Delhi

Cambridge University Press
The Edinburgh Building, Cambridge CB2 8RU, UK

Published in the United States of America by Cambridge University Press, New York

www.cambridge.org
Information on this title: www.cambridge.org/9780521073691

First published 1969
This digitally printed version 2008

A catalogue record for this publication is available from the British Library

Library of Congress Catalogue Card Number: 69–10571

ISBN 978-0-521-07369-1 hardback
ISBN 978-0-521-09336-1 paperback

Contents

	Preface	*page* vii
1.	INTRODUCTORY	1
2.	FREQUENCY DISTRIBUTIONS	4
3.	METHOD OF MOMENTS	12
4.	PEARSON'S SYSTEM OF FREQUENCY CURVES	35
5.	CALCULATION	47
6.	CURVES REPRESENTED BY SERIES EXPANSIONS	110
7.	TRANSLATION SYSTEMS	122
8.	FREQUENCY SURFACES	135
9.	STANDARD ERRORS	151
10.	TESTS OF GOODNESS OF FIT	165

Appendixes

1.	CORRECTIONS FOR MOMENTS	177
2.	B AND Γ FUNCTIONS	182
3.	THE EQUATION TO THE NORMAL SURFACE	185
4.	THE INTEGRATION OF SOME EXPRESSIONS CONNECTED WITH THE NORMAL CURVE OF ERROR	188
5.	OTHER METHODS OF FITTING CURVES	193
6.	KEY TO THE ACTUARIAL TERMS AND SYMBOLS USED	198
7.	ABRIDGED READING	201
8.	REFERENCES AND BIBLIOGRAPHY	202
9.	TABLES	206
	INDEX	214

Preface

Sir William Elderton's *Frequency Curves and Correlation* was originally published in 1906. It has remained a widely used book for upwards of sixty years, and has appeared in four editions. With each new edition the author took the opportunity to up-date the material in his book. With the passage of time, however, it became clear that further, and substantial, revision would be needed to include an adequate treatment of some more recent developments, and remove some material which was no longer of the same relative importance under present-day conditions.

The present work is the result of this revision of the fourth edition. The revision has been sufficiently drastic to warrant it being regarded as a new book, but much of the original material is still included, so that one might also consider it as a kind of 'fifth edition'.

Although there are substantial changes from the fourth edition of *Frequency Curves and Correlation*, especial attention has been given to retaining the original character of the work. The new discussion of curves other than those in the Pearson system has been presented in a form as similar as possible to that in the earlier editions. Further, the level of treatment is as close as possible to that in the original work.

Practically all the material on correlation has been removed. This subject is now covered quite adequately in many textbooks. Of recent years it is the lucid description of Pearson system curves (in particular) which has underlain the continued popularity of Sir William Elderton's book. For this reason, the space freed by omission of material on correlation has been used in an attempt to provide comparable treatment of other distributions, including some multivariate distributions.

While care has been taken to retain, as far as possible, the tone of the original—and even extend it to the newly introduced material—historical changes over more than sixty years have rendered some amendment of the original text desirable. Such amendments have been applied sparingly, without altering the original meaning in matters of opinion except where later research has changed the data on which opinion is formed. Sir William Elderton wrote, occasionally, in the first person. Where such sentences remain, it is to be understood that the personal pronoun does not apply to the present author. (In the newly added material the first person is not used.)

The methods of calculation adumbrated in the original work are those appropriate to desk calculators. With the advent of much more efficient forms of calculating assistance these methods may be found less important. They are still, however, of interest and can be useful on occasion. Their general description has, therefore, been retained, though detailed discussion of operations with arithmometers has been omitted.

Chapel Hill, North Carolina N. L. J.

1. Introductory

1. The ordinary treatment of probability begins with the assumption that the chance that a certain event will occur is known, and proceeds to solve the problems that arise from the combination of events or the repetition of a particular experiment; it proves that a certain result is more likely to occur from experiment than any other, that a result based on a limited number of trials is unlikely to differ greatly from the expected result, and that the proportional deviation from the most probable result will generally decrease as the number of trials is increased.

Experiments can easily be made to show that the theoretical method leads to results which can be realized in practice when the probabilities can be estimated accurately beforehand; for example, various trials have been made with coin tossing in which it has been found that if five coins are tossed together and the number of them coming down 'heads' is recorded, then the distribution of the cases will agree with the binomial expansion $(\frac{1}{2}+\frac{1}{2})^5$ as the ordinary theory leads us to expect. Sequences of 'heads' or 'tails' form a series approximating to the geometrical progression with a common ratio of $\frac{1}{2}$, and the drawing of cards from a pack gives a result closely agreeing with the numbers that theoretical work suggests.

2. It frequently happens, however, that the probabilities are not known, and it is impossible to tell whether we are dealing with an experiment like coin tossing or sequences or card-drawing; in fact, the only thing known is the distribution of the number of cases into certain groups, and in these circumstances the inverse problem of tracing the theoretical series to which the statistics approximate may become an important matter. The difficulty of the subject is increased because statistics do not give the theoretical distribution exactly, and it is impossible to tell where the differences between the actual and theoretical results lie. To make the position clearer it will be well to restate the problem and ask whether it is possible to find the theoretical series to which a series, resulting from a statistical experiment, approximates. It may be difficult, perhaps impossible, to trace the probabilities corresponding to a given case, but yet practicable to form a reasonable opinion of the series of numbers that might be reached if the experiment could be repeated an infinite number of times. On turning to the reasons which make it advisable to find this ideal result to which statistics approach, it will be seen that the exact elementary probabilities are not of supreme importance, and a reasonable representation of the series is of far greater practical value. We notice that one of the first objects of a statistician or an actuary dealing with statistical work is to express the observations in a simple form so that practical conclusions can be easily drawn from the figures that have been collected. If the available statistics fall naturally into fifty or sixty groups, he has to decide how they can be arranged to bring out the important features of the problem on which he is working; whereas if he can find a few numbers closely connected with the original series which can be used as an index to the whole, he can then give the result in a way that might assist comparison with similar statistics, and enable others who have to deal with the facts to appreciate the whole distribution more readily than they could do if it remained in its original form. The statistician has also to supply approximate values for intermediate terms when only a few can be obtained from his experience, or complete or continue a series when only a part of it is known. In many cases he has to keep the same terms as in his original series, but remove the roughnesses of material due to limitations in the number of cases available for his investigation; that is, he has to graduate his data.

3. In reality these objects are much alike, for if the statistical tables can be represented by an algebraic or transcendental formula, we can replace the whole series of numbers by a few values (the constants in the formula) which, if we deal systematically with the distributions we meet, facilitate comparison or enable us to supply missing terms, while the roughness of the original material can be removed by making a suitable formula represent the original statistics as nearly as possible. If a formula is based on theoretical considerations, it may also give a solution of the problem in probabilities mentioned at the outset, and we see that both the practical and theoretical requirements can be dealt with at the same time, for the smooth series sought by the theoretical student is the same thing as the formula required for practical work.

4. The advantages of any system of curves depend on the simplicity of the formulae and the number of classes of observations that can be dealt with satisfactorily, for a complicated expression is very little improvement on the original groups of statistics, and a system which is not capable of general application leaves the statistician in difficulties whenever it breaks down. One other thing is necessary; if a formula is known to be a suitable one, there must be some method of finding the arithmetical constants that will give a good agreement in the particular case. Such a method, if it is to be of practical use, must be simple, reliable and capable of general and systematic application.

A broad idea of the objects to be accomplished ought to be kept clearly before the mind; they are likely to be forgotten because of the large amount of detail necessarily connected with the subject. It is also important because the advantages of systematic treatment are often overlooked, and short cuts and rough and ready methods are adopted to the detriment of the work, and formulae having no scientific basis and having no connection with others suitable to similar cases are sometimes used in rather haphazard fashion by statisticians. The consequence is that generalization is impossible, and where a law might be found one can see little but a great variety of attempts by energetic workers to reach their own conclusions regardless of the value of comparative statistics.

2. Frequency distributions

1. If statistics are arranged so as to show the number of times, or frequency with which, an event happens in a particular way, then the arrangement is a frequency distribution. Although some of our results will be of wider applicability, we shall generally confine our attention to these distributions.

It is necessary to have a name for the formula used to describe such distributions, and the term 'frequency-curve' has been adopted for the purpose.

2. Some distributions give the number of cases falling in a certain group of values of the independent variable, while others (e.g. Example 5 of Table 1) give the number of cases for an exact value. In the former case the exact values of the independent variable to which the groups correspond must be considered; for instance, 'exposed to risk at age x' includes those from $x-\frac{1}{2}$ to $x+\frac{1}{2}$, but the number of deaths at duration n those from n to $n+1$. When statistics are represented graphically, effect should be given to these differences, and, to bring out the points a little more clearly, the diagrams on pages 6 and 7 have been prepared. Note that 'curtate duration' in Example 1 is naturally represented by a frequency polygon, as it takes only integer values, but if the data are regarded as giving actual duration they should be represented by a histogram, since generally 'curtate duration n' means 'actual duration n to $(n+1)$'. The drawings of distributions, such as those in the diagrams, are called frequency polygons or histograms (Examples 1, 3, 5).

3. When statistics give the number of cases for an exact value of the independent variable, it is simple to plot them in a diagram by drawing ordinates and joining their tops (frequency polygons). In the case of groups of values there is a little complication, for we can either draw a rectangle standing on the entire base (histograms) or put in ordinates at the middle points of the bases and then join their tops (Example 3). The former method gives the correct idea of the amount of information conveyed by the statistics, but, for some purposes (e.g. for seeing the possible shape of the curve), the latter is more convenient, though it is open to technical objection. Cases such as Examples 1 and 4 are best expressed by the kind of drawing given, while Example 3 though open to technical objection gives a better indication to most people of the shape of the actual distribution than a block diagram.

TABLE 1

Example 1		Example 2		Example 3	Example 4	Example 5	
Curtate durations	Withdrawals with monthly incidence '0' in year of exit, *Principles and Methods* (p. 92)	Ages	Exposed to risk of sickness (Watson, *M.U. Tables*, p. 19)	Existing at close of observations Without Profit 'Old' Assurances	Existing at close of observations 'Old' Annuities (females)	Terms of the expansion of $1000(\frac{3}{4}+\frac{1}{4})^{12}$	No. of term
1	308	–19	34	...	...	32	1
2	200	20–24	145	...	...	127	2
3	118	25–29	156	...	...	232	3
4	69	30–34	145	...	...	258	4
5	59	35–39	123	...	...	194	5
6	44	40–44	103	3	...	103	6
7	29	45–49	86	9	...	40	7
8	28	50–54	71	42	...	11	8
9	26	55–59	55	111	29	2	9
10	21	60–64	37	176	23	1	10
11	18	65–69	21	200	81	...	11
12	18	70–74	13	193	151	...	...
13	12	75–79	7	160	192	...	...
14	11	80–84	3	73	239	...	...
15	5	85–89	1	26	157	...	...
16	11	90–94	...	6	93	...	...
17	7	95–99	...	1	29	...	...
18	6	100–	...	...	6	...	...
19	1	...	...	...	...	...	...
20	3	...	...	...	...	...	...
21	1	...	...	...	...	...	...
22	3	...	...	...	...	...	...
23	2	...	...	...	...	...	...
...	1000	...	1000	1000	1000	1000	...
True total	1308	...	2 995 724	2674	172	...	...
Mean	4·182	...	37·8750	68·485	79·400	3·998	...
Standard deviation	4·1996	...	2·76810	1·771288	1·774894	1·46215	...
Type	I	...	I	II	VII	...	...

4. The reader is no doubt already familiar with the fact that statistics tend towards a smooth series as the total number of cases is increased, and from this it can be seen how naturally practical statistics lead to the conception of a frequency-curve to describe the smooth distribution that would be obtained if an infinite supply of homogeneous material were available for investigation. In other words, such curves would give an approximation to the total 'population' of which the particular case investigated was a sample.

5. It may be noticed that a frequency-curve can be interpreted to give a frequency corresponding to every value of the independent variable along the whole range of the distribution, and will not restrict us to a few more or less arbitrary groups as is necessary with

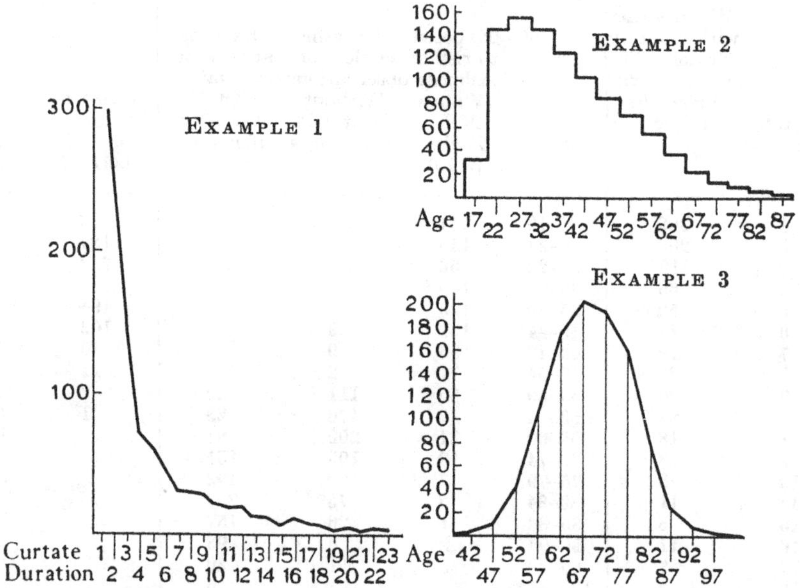

actual statistics. The binomial series and geometrical progression do the same when we imagine we are dealing with something that can be divided into a very large number of groups. Thus, if we mix a large quantity of sand of two colours and take out a fixed quantity of the mixture and record the number of grains of sand of either colour in each drawing, we should obtain a continuous curve from a large number of trials.

6. We will now define some important functions. When a distribution is arranged according to the progressive values of a variable characteristic, e.g. duration, age, etc., the average value of that characteristic (not the average of the frequencies) is called the *mean* of the distribution, and is given by

$$\frac{f_a \times a + f_b \times b + f_c \times c + \ldots + f_n \times n}{f_a + f_b + f_c + \ldots + f_n}$$

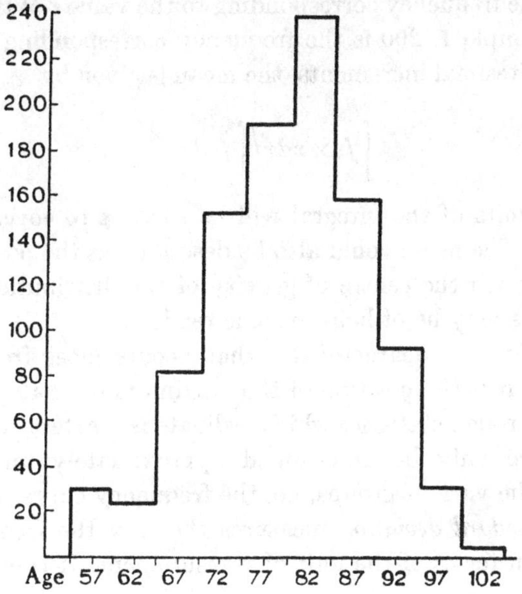

EXAMPLE 4

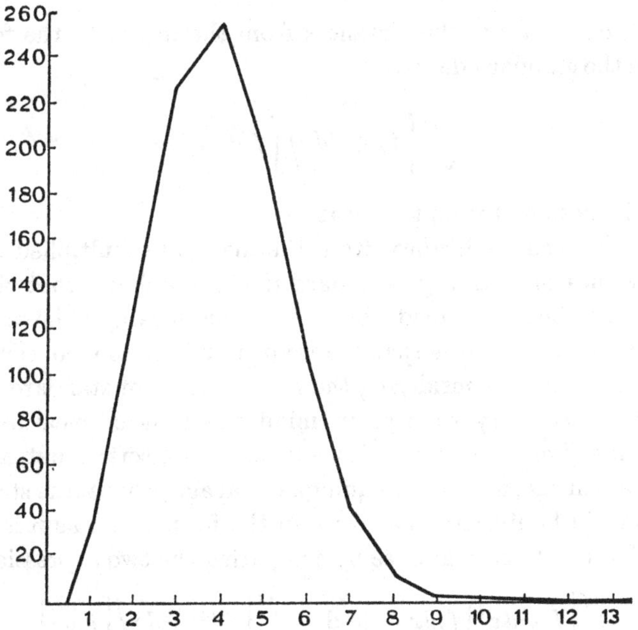

EXAMPLE 5

where f_r is the frequency corresponding to the value r of the variable; thus, in Example 1, 200 is the frequency corresponding to 2. If we assume infinitesimal increments, the mean is given by

$$\int f_x \times x\,dx \Big/ \int f_x\,dx$$

where the limits of the integral will be such as to cover the whole distribution. The mean could also be described as the position of the ordinate through the centre of gravity of the distribution (centroid vertical); this may be of help to some readers.

The *mode* is the characteristic that occurs most frequently; in other words, it is the position of the maximum ordinate. We cannot tell from the rough statistics which ordinate is greatest and the mode can therefore only be determined approximately until the law connecting the various groups, i.e. the frequency-curve, is known.

7. The *standard deviation* measures the way the frequencies are distributed in terms of the unit of measurement. It is given by

$$\sqrt{\left\{\frac{f_a a'^2 + f_b b'^2 + \ldots + f_n n'^2}{f_a + f_b + \ldots + f_n}\right\}}$$

where a', b', ... n' are the distances from the mean. In the form of integrals the standard deviation is

$$\sqrt{\left\{\int f_x \times x^2\,dx \Big/ \int f_x\,dx\right\}}$$

where x is measured from the mean.

As the frequencies farthest from the mean are multiplied by the largest values of x, a large standard deviation shows that the frequency distribution spreads out from the mean, while a small standard deviation shows that the frequency is closely concentrated about the mean. In considering the relative sizes of standard deviations, it is necessary to bear in mind the unit of measurement, because, if a given distribution is arranged in two series, first, according to years of age, and then in quinquennial age groups, the standard deviation will be five times as large in the former case as it is in the latter. This can be seen at once by comparing the two expressions

$$\sqrt{\left\{\int f_x x^2\,dx \Big/ \int f_x\,dx\right\}} \quad \text{and} \quad \sqrt{\left\{\int f_x (\tfrac{1}{5}x)^2\,dx \Big/ \int f_x\,dx\right\}}$$

The former is obviously five times the latter. The values of the standard deviations are given in Table 1 for each case. The diagram on page 10 shows two curves having the same mean B and approximately the same area, but the dotted curve has the larger standard deviation because it spreads out more on each side of the mean.

The *mean deviation* is the average of the absolute values of deviations from the mean. It is given by

$$(f_a|a'|+f_b|b'|+\dots+f_n|n'|)/(f_a+f_b+\dots+f_n)$$

or, in the form of integrals by

$$\int f_x|x|\,dx \Big/ \int f_x\,dx$$

where x is measured from the mean.

The reader will notice from the algebraic expressions given above that the mean, mode, standard deviation and mean deviation are not dependent on the number of cases (i.e. on the absolute size of the curve), but merely on the way they are distributed (i.e. on the proportionate numbers or the shape of the curve). The standard deviation measures the 'spread' or 'scatter' of the statistics from the mean. Its square is called the *variance*, and is often denoted var (x), x being the variable concerned.

8. An examination of frequency distributions (see Table 1 and pp. 6 and 7) shows that most of them start at zero, gradually rise to a maximum, and then fall sometimes at a very different rate. If the rise and fall are at the same rate, distribution will be symmetrical about the mean, which must then coincide with the mode. The difference between the mean and mode is therefore a function of the *skewness* or deviation from symmetry. In order to get a satisfactory measure, the spread of the material must be taken into account, and this leads us to measure skewness by the distance between mean and mode divided by standard deviation. If the mean is on the left-hand side of the mode when the statistics are plotted out in diagram, this function will be negative, and to remember the sign it is convenient to write:

$$\text{Skewness} = \frac{\text{Mean}-\text{Mode}}{\text{S.D.}}$$

The diagram on page 10 will help to show the rationale of the measure for skewness. It gives two curves having the same mean B

and the same mode A, but with different standard deviations, and it is clear that the dotted curve, with its larger standard deviation, is more nearly symmetrical than the other curve. There are other measures of skewness, such as $\sqrt{\beta_1}$ (to be defined in section 4 of the next chapter).

9. We may summarize these functions by saying that the mean and mode fix the position of the curve on the axis; the standard deviation shows how the material is distributed about the mean, and the skewness shows the amount of the deviation from symmetry exhibited by the material.

These preliminary definitions will be sufficient for our present purpose, but the functions defined will be more easily understood when their actual connection with the practical work of curve-fitting has been studied. A student working at the subject for the first time should plot out several distributions on cross-ruled paper, in order to familiarize himself with their nature and appearance. He should also calculate means, standard deviations and mean deviations.

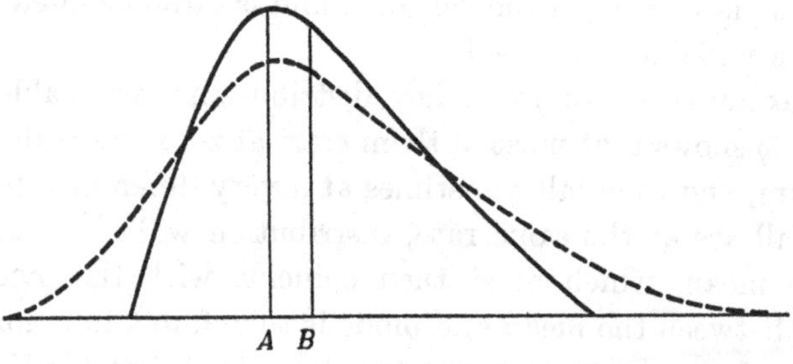

10. Up to this point we have defined our statistics as frequencies, that is, as a number of cases grouped together as alike either because they are actually alike in the sense of Example 5 or because the statistics throw them up in comparatively narrow groupings as in Examples 2, 3 and 4. When, however, we are tabulating our experience we have to deal with individual observations and they are grouped subsequently. From this point of view if there are N observations we may call them $o_1, o_2, o_3, \ldots o_N$, where o_1 may stand for the first observation and may be (see Example 3) one of the 200 existing in the 65–69 group. It might be a case 'existing' at age

66·12, and o_2 might be 'existing' at age 73·72, o_3 at 42·26, o_4 at 67·37 and so on. Then the mean is

$$\frac{1}{N}\sum_{i=1}^{N} o_i$$

We will, in fact, use the symbol N to denote the total area under a frequency curve (or volume under a surface) throughout this book, so that, for example

$$\int f_x \, dx = N.$$

3. Method of moments

1. Before we proceed to deal with suitable forms for use as frequency-curves, it will be well to see if some method of applying them to statistical examples can be found, for it is clearly useless to suggest a curve and have no way of using it. We require, therefore, a general method by which a given formula can be fitted to a particular statistical experience, and may be applied to any expression (for instance, Makeham's formula for the force of mortality) on which we may have decided as the basis of graduation. The first point to be noticed in searching for a method is that if there are n constants in the formula, we must form n equations between the formula and the statistics. Thus, if we have three terms, say, $y = 20$, 40, and 88, when $x = 1$, 2, and 3 respectively, and wish to use the curve $y = a+bx+cx^2$ to describe them, we can, of course, find values of a, b and c so that each item is exactly reproduced by equating as follows:

$$a+b+c = 20$$
$$a+2b+2^2c = 40$$
$$a+3b+3^2c = 88$$

But if we have a fourth term $y = 96$ when $x = 4$, and use the values of a, b and c found from the three equations just given, we should find that when $x = 4$, $y = 164$. This suggests that when there are more terms in the statistics than there are constants, the equations must be formed by using all the terms, not by selecting from them. The graduating curve will not necessarily reproduce exactly any of the observations, but will run evenly through the roughnesses of the observed facts so as to represent their general trend.

2. Let $a_1, a_2, a_3, \ldots a_n$ be n terms to be graduated; then, if the series were perfectly smooth and followed a known law, each term could be reproduced exactly by, say, b_1, b_2, b_3, b_n, where $a_1 = b_1$, $a_2 = b_2$, $a_3 = b_3$, ... and $a_n = b_n$. Now, if we consider the two series (the a's and the b's), we see that since each term is reproduced exactly

$$\sum_{r=1}^{n} a_r = \sum_{r=1}^{n} b_r \quad \text{and} \quad \sum_{r=1}^{n} c_r a_r = \sum_{r=1}^{n} c_r b_r$$

where c_r is a numerical coefficient.

This suggests a possible method to apply when each term cannot be reproduced exactly. The total of the graduated figures must be made equal to the total of the ungraduated, and the further equations necessary for finding the unknown constants must be formed by multiplying the various terms by different factors and similarly equating the sums of the graduated and ungraduated products, i.e. $\sum c_r a_r = \sum c_r b_r$. It still remains to decide the best form to be given to c_r, and the mean being equal to

$$\frac{a_1+2a_2+\ldots+na_n}{a_1+a_2+\ldots+a_n}$$

suggests that $c_r = r$ should give one reasonable equation. Again, since we shall have to use some function of r which, when applied to the graduation formula, will give an integrable form (otherwise we cannot make an equation between $\sum c_r b_r$ and $\sum c_r a_r$), the powers of r suggest themselves as convenient when integration by parts is attempted. If, therefore, we write $c_r = r^t$ and give t successively the values 0, 1, 2, ..., we can obtain as many equations as we require, and from the first two of them we find, successively, the area and the mean, which will be the same in the graduated and ungraduated figures.

This method is known as the Method of Moments (cf. moments of inertia), and experience has shown that it is a satisfactory method of fitting a curve to an actual statistical experience. Some other methods are described in Appendix 5.

3. Applying the method to the data given above ($x = 1$, $y = 20$; $x = 2$, $y = 40$; $x = 3$, $y = 88$; $x = 4$, $y = 96$), we have

$$(a+b+c)+(a+2b+2^2c)+(a+3b+3^2c)+(a+4b+4^2c) = 20+40+88+96$$

$$(a+b+c)+2(a+2b+2^2c)+3(a+3b+3^2c)+4(a+4b+4^2c) = 20+2\times 40+3\times 88+4\times 96$$

$$(a+b+c)+2^2(a+2b+2^2c)+3^2(a+3b+3^2c)+4^2(a+4b+4^2c) = 20+2^2\times 40+3^2\times 88+4^2\times 96$$

or

$$\begin{aligned} 4a+\ 10b+\ 30c &= 244 \\ 10a+\ 30b+100c &= 748 \\ 30a+100b+354c &= 2508 \end{aligned}$$

The solution of these equations gives

$$a = -23{\cdot}0$$
$$b = \quad 42.6$$
$$c = -\ 3{\cdot}0$$

or

$$x = 1 \qquad y = 16{\cdot}6$$
$$x = 2 \qquad y = 50{\cdot}2$$
$$x = 3 \qquad y = 77{\cdot}8$$
$$x = 4 \qquad y = 99{\cdot}4$$

This is a very simple example, but it will probably help to show the way results are reached, and will serve as a foundation for what follows.

4. The nth moment of a particular frequency is defined as the product of the frequency and the nth power of the distance of the frequency from the vertical about which moments are being taken; or the nth moment of any *ordinate* y of a frequency-curve about the vertical through a point distance x from it, is yx^n, and the nth moment of the whole distribution treated as a series of ordinates is $y_1x_1^n+y_2x_2^n+\ldots$, where $y_1+y_2+\ldots$ is the total frequency. Thus, in Example 5, the third moment of the frequency 40 for term 7 about the vertical through 3 is $40\times(+4)^3$.

5. If the ordinates are known, we can calculate the moment for them immediately by multiplying the frequencies by the powers of the distances between them and the vertical about which the moments are required and then adding the results, care being taken to give the distances their proper signs. If areas are given, an approximation is made by assuming them to be concentrated about the ordinates at the middle points of the bases on which they stand. The columns after the third in Table 2 show the calculation of the moments about the vertical through age 77 for Example 4 of Table 1, on the assumption that the frequencies are concentrated at the middle points of the bases.

The unit of grouping has been taken as 5 years, and if, as is often convenient, we assume the total frequency to be unity, the totals will have to be divided by 1000. We should generally deal with the actual numbers that occur, but as they have been given in Table 1 as the distribution of 1000 cases, it will be better to use them in that way in the present case. The numbers -4, -3, ... in col. (3) show the distances from age 77 in terms of the unit of grouping. The centre of any other group would have done almost as well as 77; it is convenient to choose the arbitrary origin so that it is near the mean of the distribution. This makes easier the calculation of the moments about the mean (a result frequently required), and enables the calculator to get a rough check on these moments by comparing them with those about the arbitrary origin. The cols. (4)–(7) are sufficiently

TABLE 2

Central age of group x (1)	Frequency f (2)	$(x-77)/5 = s$ (3)	$f\times s$ (4)	$f\times s^2$ (5)	$f\times s^3$ (6)	$f\times s^4$ (7)
57	29	−4	116	464	1856	7424
62	23	−3	69	207	621	1863
67	81	−2	162	324	648	1296
72	151	−1	151	151	151	151
77	192	0	−498	...	−3276	...
82	239	1	239	239	239	239
87	157	2	314	628	1256	2512
92	93	3	279	837	2511	7533
97	29	4	116	464	1856	7424
102	6	5	30	150	750	3750
Totals	1000	...	+978	3464	+6612	32 192
			+480		+3336	

NOTATION FOR MOMENTS

N = total frequency.
ν_n = nth unadjusted statistical moment about mean.
ν'_n = nth unadjusted statistical moment about any other point.
μ_n = nth moment from curve about mean.
= nth adjusted statistical moment about mean.
μ'_n = nth moment from curve about other point.
= nth adjusted statistical moment about other point.

NOTE.—ν, ν', μ and μ' always refer to a total frequency of unity.

explained by their headings; they are formed successively and checked by multiplying f by s^4, the values of s^4 being taken from a table of the powers of the natural numbers.

The arithmetical work may be checked in other ways; for instance, instead of checking the final column by multiplying each term by the appropriate value of x^4, we can form a new column $(x+1)^4f$, which is the same thing as

$$x^4f+4x^3f+6x^2f+4xf+f$$

The total of this new column can therefore be used to give a check on the multiplication and addition. In the numerical example (Table 2) we should have $29 \times (-3)^4$; $23 \times (-2)^4$, etc., 6×6^4; the total of such a column is 69 240, which agrees with the totals of cols. (2)–(7) in the following way:

$$\begin{aligned} & 32\,192 \\ 4 \times 3336 = {} & 13\,344 \\ 6 \times 3464 = {} & 20\,784 \\ 4 \times 480 = {} & 1\,920 \\ & 1\,000 \\ \hline & 69\,240 \end{aligned}$$

Helpful tables (Powers and Fourth moments) will be found in *Biometrika Tables for Statisticians* edited by E. S. Pearson and H. O. Hartley (Cambridge University Press). I shall in future refer to this book as *Biometrika Tables.* A student can manage without this volume, but at some expense of trouble.

In earlier editions of this book, *Tables for Statisticians and Biometricians* (2 volumes) edited by K. Pearson (Cambridge University Press) was used as a reference. *Biometrika Tables* contains most tables which were in the first volume of the earlier work (now out of print). Where this is not so, the abbreviated reference *Tables for Statisticians* will be used.

6. It has so far been assumed that moments can be calculated about any point, but it is frequently inconvenient to do so; for if we had required them about age 79·4, we should have had to multiply by powers of (57—79·4)/5, of (62—79·4)/5 and so on, and it is quite clear that the labour would have been very great. In such a case we can, however, take the moments about any other more convenient point, and then modify them in the following way:

Let the distance between A, about which the moments are known, and B, about which they are required, be $+d$; thus, if we want moments about 25·7 and have found them about 25, d is 0·7; if we had found them about 26, d would have been —0·3.

Then, if the distance of any ordinate y_r from A is X_r, and from B is x_r, then

$$x_r = X_r - d \quad \text{and} \quad x_r^n = (X_r - d)^n$$

Now, the nth moment of the whole distribution treated as a series of ordinates is $\sum y_r X_r^n$ about A, and $\sum y_r x_r^n$ about B; so we have

$$\begin{aligned}\nu_n'' &= \sum y_r x_r^n = \sum y_r(X_r-d)^n \\ &= \sum[y_r\{X_r^n - ndX_r^{n-1} + \dots + (-1)^n d^n\}] \\ &= \nu_n' - nd\nu_{n-1}' + \frac{n(n-1)}{2!}d^2\nu_{n-2}' - \dots \qquad \dots\dots(1)\end{aligned}$$

where ν_n'' is written for the nth moment about B, and ν_n' the nth moment about A.

Instead of (1) we may proceed as follows:

$$\begin{aligned}\nu_n' &= \sum y_r X_r^n = \sum[y_r(x_r+d)^n] \\ &= \nu_n'' + nd\nu_{n-1}'' + \frac{n(n-1)}{2!}d^2\nu_{n-2}'' + \dots\end{aligned}$$

$$\therefore \qquad \nu_n'' = \nu_n' - nd\nu_{n-1}'' - \frac{n(n-1)}{2!}d^2\nu_{n-2}'' - \dots \qquad \dots\dots(2)$$

There is little to choose between these two formulae, and of course they give identical results.

7. We will now apply them to work out the moments about the centroid vertical (i.e. vertical through the mean) for the example in Table 2. The distance of the mean from any point is

$$\frac{\sum(X_r y_r)}{\sum y_r} = \frac{\sum(X_r y_r)}{N}$$

where N is the total frequency; or we may say that the distance of the mean from any point is the first moment of the distribution about the vertical through that point. It follows that the first moment about the centroid vertical is zero, so that if such moments are required the term involving ν_1'' in (2) is zero. When we deal with frequency-curves we shall see that we generally require moments about the centroid vertical and in designating them we shall leave out the dashes and use ν.

8. The arithmetical work is as follows:

The totals in cols. (4)–(7) are divided by the number of observations (total of col. (2)), and the quotients are the moments (ν') about 77. The moments are dealt with as having reference to a case where unity is the total frequency, i.e. proportional, not actual, frequencies are dealt with.

$$\nu_1' = 0{\cdot}480 \qquad \nu_2' = 3{\cdot}464$$

$$\nu_3' = 3{\cdot}336 \qquad \nu_4' = 32{\cdot}192$$

The value of ν_1' gives the mean age $= 77+5\times 0{\cdot}480 = 79{\cdot}4$. In order to use formula (1) or (2), the value of d is required, and when the calculation of moments has to be made about the centroid vertical its value is, as we have seen above, the same as ν_1'; in the present case it is the first moment about the vertical through age 77. The powers of d are next calculated by logarithms. As it happens d is a comparatively simple number; if it had been 0·48327, say, the propriety of using logarithms would have been more obvious

$$d^2 = 0{\cdot}2304 \qquad d^3 = +0{\cdot}110592 \qquad d^4 = 0{\cdot}0530842$$

Remembering that ν_1 is zero and ν_0' and ν_0 are each unity because they are the total frequency, we reach

$$\nu_2 = 3{\cdot}2336 \qquad \nu_3 = -1{\cdot}430976 \qquad \nu_4 = 30{\cdot}416289$$

It is wise to work to a large number of decimal places because, owing to the subtractions involved, calculations which began with, say, seven figures may end with only five. It is well, therefore, to use a seven-place logarithm table (e.g. Chambers's) and antilogarithm table if mechanical aids are not available.

It will be noticed that the terms required in the calculation of successive moments can be formed continuously. Thus, in formula (1), we require to calculate the following multiples of ν_2':

$$\begin{aligned} &1 \text{ for the second moment}\\ &3d \text{ for the third moment}\\ &\frac{4\,.\,3}{2}d^2 \text{ for the fourth moment.}\end{aligned}$$

I have found it convenient to adopt a regular system in calculating moments, as in other statistical work, and create the habit of putting results and calculations in fixed positions, so that the arithmetic, which is sometimes complicated, can be followed quickly and can be confirmed or rectified more easily.

9. Although the above is the usual way to calculate moments, another method was suggested by the late Sir G. F. Hardy and used by him in his graduation of the British Offices Tables 1863–93. He pointed out that by summing the statistical numbers and forming a new series in the same way as is done by actuaries when the N_x column is formed from the D_x column and then summing these results (cf. the S column), and so on, equations can be formed which give the same results as the method of moments. The arrangement in Table 3 shows both the method of calculation and the form of the expression obtained by the process.

TABLE 3

No. of term	Function	Sum of function	Second sum of function	Third sum of function
1	$f(1)$	$f(1)+f(2)+\ldots+f(n)$	$f(1)+2f(2)+\ldots+nf(n)$	$f(1)+3f(2)+6f(3)+\ldots+\frac{n(n+1)}{2}f(n)$
2	$f(2)$	$f(2)+f(3)+\ldots+f(n)$	$f(2)+2f(3)+\ldots+(n-1)f(n)$	$f(2)+3f(3)+\ldots+\frac{(n-1)n}{2}f(n)$
3	$f(3)$	$f(3)+\ldots+f(n)$	$f(3)+2f(4)+\ldots+(n-2)f(n)$	$f(3)+3f(4)+\ldots+\frac{(n-2)(n-1)}{2}f(n)$
4	$f(4)$	$f(4)+\ldots+f(n)$	$f(4)+2f(5)+\ldots+(n-3)f(n)$	$f(4)+3f(5)+\ldots+\frac{(n-3)(n-2)}{2}f(n)$
.	.	.	.	.
.	.	.	.	.
.	.	.	.	.
$n-2$	$f(n-2)$	$f(n-2)+f(n-1)+f(n)$	$f(n-2)+2f(n-1)+3f(n$	$f(n-2)+3f(n-1)+6f(n)$
$n-1$	$f(n-1)$	$f(n-1)+f(n)$	$f(n-1)+2f(n)$	$f(n-1)+3f(n)$
n	$f(n)$	$f(n)$	$f(n)$	$f(n)$
General form of the sums on line 1		$\sum_{n=1}^{n} f(n) = \nu_0'$	$\sum_{n=1}^{n} nf(n) = \nu_1'$	$\sum_{n=1}^{n} \frac{n^2+n}{2} f(n) = \frac{1}{2}(\nu_2'+\nu_1')$

Fourth sum of function	Fifth sum of function
$f(1)+4f(2)+\ldots+\frac{n(n+1)(n+2)}{3!}f(n)$	$f(1)+5f(2)+\ldots+\frac{n(n+1)(n+2)(n+3)}{4!}f(n)$
$f(2)+4f(3)+\ldots+\frac{(n-1)n(n+1)}{3!}f(n)$	$f(2)+5f(3)+\ldots+\frac{(n-1)n(n+1)(n+2)}{4!}(fn)$
$f(3)+4f(4)+\ldots+\frac{(n-2)(n-1)n}{3!}f(n)$	$f(3)+5f(4)+\ldots+\frac{(n-2)(n-1)n(n+1)}{4!}f(n)$
$f(4)+4f(5)+\ldots+\frac{(n-3)(n-2)(n-1)}{3!}f(n)$	$f(4)+5f(5)+\ldots+\frac{(n-3)(n-2)(n-1)n}{4!}f(n)$
.	.
.	.
.	.
$f(n-2)+4f(n-1)+10f(n)$	$f(n-2)+5f(n-1)+15f(n)$
$f(n-1)+4f(n)$	$f(n-1)+5f(n)$
$f(n)$	$f(n)$
$\sum_{n=1}^{n} \frac{n^3+3n^2+2n}{6} f(n) = \frac{1}{6}(\nu_3'+3\nu_2'+2\nu_1')$	$\sum_{n=1}^{n} \frac{n^4+6n^3+11n^2+6n}{24} f(n) = \frac{1}{24}(\nu_4'+6\nu_3'+11\nu_2'+6\nu_1')$

Considering the line opposite the first term, we notice that the sum of the series is given, and that the second summation, which we will call S_2 when the total frequency is taken as unity, gives the first moment of the whole distribution about a vertical situated at unit distance before the point corresponding to $f(1)$. Still considering only the first line, we see that S_3 gives each function multiplied by coefficients of the form $n(n+1)/2!$ or $(n^2+n)/2!$, i.e. it gives $(\nu_2'+\nu_1')/2$, where ν' is written for the moment, because by definition the tth moment (ν_t') of the whole distribution is given by the sum of $n^t f(n)$ for all values of n. S_4 and S_5 give each function multiplied by $(n^3+3n^2+2n)/6$ and $(n^4+6n^3+11n^2+6n)/24$ respectively.

The following equations result:

$$S_2 = \nu_1' \qquad S_4 = \tfrac{1}{6}(\nu_3'+3\nu_2'+2\nu_1')$$

$$S_3 = \tfrac{1}{2}(\nu_2'+\nu_1') \qquad S_5 = \tfrac{1}{24}(\nu_4'+6\nu_3'+11\nu_2'+6\nu_1')$$

These equations enable us to calculate the moments about the selected origin, but if it is necessary to find moments about the mean, the following relations are more convenient; they can be reached by substituting in the above the values in formula (2), and remembering that $S_2 = d$.

$$\nu_2 = 2S_3-d(1+d)$$

$$\nu_3 = 6S_4-3\nu_2(1+d)-d(1+d)(2+d)$$

$$\nu_4 = 24S_5-2\nu_3\{2(1+d)+1\}-\nu_2\{6(1+d)(2+d)-1\}-d(1+d)(2+d)(3+d)$$

10. Table 4 shows the working in the numerical example already dealt with by the direct method. The fifth sum is unnecessary, as the total of the items in the fourth sum gives the only value required:

From the totals of the columns we have

$S_2 = d = 5{\cdot}48$ $S_3 = 19{\cdot}372$ $S_4 = 54{\cdot}508$ and $S_5 = 132{\cdot}503$

The first value S_2 or d shows that the mean is at age

$$52+5{\cdot}48\times 5 = 79{\cdot}4.$$

The age 52 is used because it is the centre of the group before that in which numbers occur, and, as has been already remarked, the summation method assumes the work to be done with reference to this position. The application of the formula for ν_2, ν_3 and ν_4, given above, enables us to find

$$\nu_2 = 3{\cdot}2336 \qquad \nu_3 = -1{\cdot}43099 \qquad \nu_4 = 30{\cdot}4164$$

11. We may save arithmetical work in several ways when using the summation method. If, instead of making all the calculations implied in Table 3, we stop at the sums next above the lines ruled in the various columns we shall have as the final totals

$$\Sigma f(n); \quad \Sigma nf(n); \quad \Sigma\frac{n(n-1)}{2!}f(n); \quad \Sigma\frac{n(n-1)(n-2)}{3!}f(n)$$

and
$$\Sigma\frac{n(n-1)(n-2)(n-3)}{4!}f(n).$$

The formulae of § 9 for S_3, S_4 and S_5 in terms of ν_1', ν_2' etc. require modification only by altering the alternate signs from + to −.

TABLE 4

Frequency	First sum	Second sum	Third sum	Fourth sum
29	1000	5 480	19 372	54 508
23	971	4 480	13 892	35 136
81	948	3 509	9 412	21 244
151	867	2 561	5 903	11 832
192	716	1 694	3 342	5 929
239	524	978	1 648	2 587
157	285	454	670	939
93	128	169	216	269
29	35	41	47	53
6	6	6	6	6
Total (for check) 1000	5480	19 372	54 508	132 503

The form of moment given in this paragraph (§ 11) has been called 'factorial moments'.* A little further saving of work can be effected by taking the figures up to the totals next below the lines ruled in Table 3. This gives the same result as that just obtained if the origin is assumed to be shifted one space. It will suffice if we take this second case for a numerical example and using the figures from Table 4, we should work only so far as 4480 for the second sum, 9412 for the third, 11 832 for the fourth and for the fifth we sum the last six entries in the fourth column and obtain 9783.

* The semi-invariants (or half-invariants) used by Thiele and other writers can be obtained from moments. The second and third semi-invariants are the same as the second and third moments about the mean and the fourth semi-invariant is the fourth moment less three times the square of the second moment ($\mu_4-3\mu_2^2$).

12. These are the direct ways of using the summation method, but, as in the multiplication method of calculating moments, we can shorten the work by using a central term instead of the first term as the starting point or arbitrary origin.* We shall now use this arrangement with the 'factorial moment' form. A little care is necessary, because, though there is no difficulty about the interpretation of the sums on the positive side of the selected point, the moments for the terms on the negative side assume that multiplications are made by the powers of negative quantities. Table 4(A) gives an example of summations that have to be made. The figure 978 is $\sum nf(n)$ for values on the positive side of the arbitrary origin and 498 is the similar sum on the negative side, ignoring sign, say $\sum mf(-m)$. The mean is found by dividing the difference, $\sum nf(n) - \sum mf(-m)$, by the total frequency, i.e. $(978-498)/1000 = 0{\cdot}48$. Taking, now, the final figures in the columns headed 'Third sum', 670 represents $\dfrac{\sum n(n-1)f(n)}{2!}$ and 324 is the corresponding figure on the negative side; similarly with the other columns.

Now reverting to the expressions in § 11, which relate only to positive summations, we can write

$$\nu_2' = 2S_3' + S_2'$$

$$\nu_3' = 6S_4' + 6S_3' + S_2'$$

$$\nu_4' = 24S_5' + 36S_4' + 14S_3' + S_2'$$

where S' is used instead of S to indicate the different system of summation.

In Table 4 (A) however we have divided the distribution into two parts, and in applying the expressions just given we must work out each separately, adding the items for even moments and subtracting for odd moments. Hence for the whole distribution

$$\nu_2' = (2\times 0{\cdot}670 + 0{\cdot}978) + (2\times 0{\cdot}324 + 0{\cdot}498) = 3{\cdot}464$$

$$\nu_3' = (6\times 0{\cdot}269 + 6\times 0{\cdot}670 + 0{\cdot}978) - (6\times 0{\cdot}139 + 6\times 0{\cdot}324 + 0{\cdot}498)$$
$$= 3{\cdot}336$$

$$\nu_4' = (24\times 0{\cdot}059 + 36\times 0{\cdot}269 + 14\times 0{\cdot}670 + 0{\cdot}978) +$$
$$(24\times 0{\cdot}029 + 36\times 0{\cdot}139 + 14\times 0{\cdot}324 + 0{\cdot}498) = 32{\cdot}194$$

and, transferring to the mean,

$$\nu_2 = 3{\cdot}2336, \quad \nu_3 = -1{\cdot}43098 \quad \text{and} \quad \nu_4 = 30{\cdot}41626.$$

* I have to thank Mr G. J. Lidstone for the suggestion of shortened summations.

We may now express the work in symbols. Writing P for summations on the positive side and N for those on the negative side of the arbitrary origin, we have

$$\nu_1' = P_2 - N_2$$

$$\nu_2' = (2P_3+P_2)+(2N_3+N_2)$$
$$= 2(P_3+N_3)+(P_2+N_2)$$

and similarly

$$\nu_3' = 6(P_4-N_4)+6(P_3-N_3)+(P_2-N_2)$$
$$\nu_4' = 24(P_5+N_5)+36(P_4+N_4)+14(P_3+N_3)+(P_2+N_2).$$

TABLE 4 (A)

Frequency	First sum	Second sum	Third sum	Fourth sum	Fifth sum
29	29	29	29	29	29
23	52	81	110	139	...
81	133	214	324	...	...
151	284	498	...	...	...
192					
239	524	978	...	...	...
157	285	454	670	...	...
93	128	169	216	269	...
29	35	41	47	53	59
6	6	6	6	6	6
1000					

13. A comparison of Table 4 (A) with Table 4 will show that a saving of numerical work is effected by using a central point as the starting point for the summation, for the sums are numerically smaller and the value of S_2 or d, which enters into the formulae on page 20, is much smaller. It will be readily appreciated that whenever there is a large number of terms the summation method, and especially the form of it given in Table 4 (A), is an improvement on the product method of calculating moments. By means of an adding machine the summations can be obtained mechanically with little trouble, even for series containing as many as a hundred terms.

14. In § 12 of Chapter 2 the mean was described alternatively in terms of the individual observations, $o_1, o_2, \ldots o_N$. Similarly the tth moment is

$$\frac{1}{N}\sum_{i=1}^{N} o_i$$

and the tth factorial moment is

$$\frac{1}{N}\sum_{i=1}^{N} o_i(o_i-1)\dots(o_i-t+1) = \frac{1}{N}\sum_{i=1}^{N} o_i^{(t)}$$

15. It is now necessary to consider the calculation of moments from the curve.

Let $y_x = f(x, a, b, c, \dots)$, where $a, b, c, \dots$ are constants to be determined.

We have seen, on pages 12 and 13, that one way of working would be to find

$$f(1, a, b, c, \dots) \times 1^n + f(2, a, b, c, \dots) \times 2^n + \dots$$

say,
$$\sum_{x=1}^{r} f(x, a, b, c, \dots) \times x^n$$

and this would give a result which might be used in forming equations if it were not for the fact that it is often impossible to find an algebraic expression for the sum of such a series in terms of the constants. It is, however, generally possible to find such an expression for the integral, and as we have defined the nth moment of an ordinate y_x as $y_x x^n$, the nth moment of the whole distribution from $x = h$ to $x = k$ is

$$\int_h^k y_x x^n \, dx \quad \text{or} \quad \int_h^k f(x, a, b, c, \dots) x^n \, dx$$

The total frequency (i.e. total number of cases investigated) is $\int_h^k y_x dx$, and the mean is $\int_h^k y_x x \, dx \Big/ \int_h^k y_x dx$, as we have already noticed.

16. If the moments from the equation to the curve are calculated in this way and equated to the moments calculated from statistics by assuming that the latter consist of a series of ordinates, an inaccuracy is introduced.

Let us consider the two cases:

(1) When the statistics are a system of isolated terms or ordinates * and we wish to pass a curve very closely through them.

(2) When they are a system of areas but the moments are calculated by assuming the areas to be concentrated at the middle points of the bases.

* Strictly speaking, not a frequency distribution but a series of values requiring graduation. This case occurs relatively seldom. Distributions have generally to be dealt with as areas for frequency-curve work because they tell the way the whole number of cases is divided in groups, and the whole area between the curve and the axis of x must therefore be used.

17. In case (1) above, the terms $y_0, y_1, y_2, \dots y_{n-1}$ are given by the statistics, and since $\int_{-\frac{1}{2}}^{\frac{1}{2}} y_x dx$ is approximately equal to y_0, it is simplest * to assume that $\int_{-\frac{1}{2}}^{n-\frac{1}{2}} y_x dx$ is given by the equation to the curve, and we have to find adjustments to counteract the error † caused by equating $\sum_{x=0}^{n-1} Xy_x$ to $\int_{-\frac{1}{2}}^{n-\frac{1}{2}} Xy_x dx$. The most practical way of overcoming the difficulty is by calculating the true area corresponding to the ordinates $y_0, y_1, \dots y_{n-1}$ by means of a quadrature formula (formula of approximate summation). Many formulae are well known, but for the present purpose it is convenient to have expressions which give approximate values of an area in terms of ordinates lying both within and without the base on which the area to be valued stands. Symbolically, these formulae express $\int_{-\frac{1}{2}}^{\frac{1}{2}} y_x dx$ in terms of $y_{-\frac{1}{2}}, y_{\frac{1}{2}}, y_{-1\frac{1}{2}}, y_{1\frac{1}{2}}$, etc., or $y_0, y_1, y_{-1}, y_2, y_{-2}$, etc.

First let

$$y_x = a+bx+cx^2+dx^3+ex^4$$

then

$$\int_{-\frac{1}{2}}^{\frac{1}{2}} y_x dx = a+\frac{c}{12}+\frac{e}{80}$$

and

$$y_0 = a$$

$$y_{-1}+y_1 = 2(a+c+e)$$

$$y_{-2}+y_2 = 2(a+4c+16e)$$

Now, assume the required integral can be equated to

$$hy_0+k(y_{-1}+y_1)+l(y_{-2}+y_2)$$

substitute the values gives just above and equate the coefficients of a, c and e respectively to 1, $\frac{1}{12}$ and $\frac{1}{80}$, and we have

$$h+2k+2l = 1$$

$$2k+8l = \tfrac{1}{12}$$

$$2k+32l = \tfrac{1}{80}$$

The solution of these equations gives

$$h = \tfrac{5178}{5760}, \qquad k = \tfrac{308}{5760} \quad \text{and} \quad l = -\tfrac{17}{5760}$$

* It is generally possible to use these limits in case (1), but if other limits have to be taken, such as 0 to n, different quadrature formulae must be used.

† Actuarial readers will notice that the error is analogous to that introduced by assuming $(1+i)^{\frac{1}{2}}a_{\overline{n|}} = \bar{a}_{\overline{n|}}$.

and we obtain

$$\int_{-\frac{1}{2}}^{\frac{1}{2}} y_x\,dx = \tfrac{1}{5760}\{5178y_0+308(y_{-1}+y_1)-17(y_{-2}+y_2)\} \qquad \text{......(1)}$$

Secondly, if

$$y_x = a+bx+cx^2+dx^3$$

$$\int_{-\frac{1}{2}}^{\frac{1}{2}} y_x\,dx = \tfrac{1}{24}\{y_{-1}+22y_0+y_1\} \qquad \text{......(2)}$$

Thirdly, if

$$y_x = a+bx+cx^2+dx^3+ex^4$$

$$\int_{-\frac{1}{2}}^{\frac{1}{2}} y_x\,dx = \tfrac{1}{1440}\{802(y_{\frac{1}{2}}+y_{-\frac{1}{2}})-93(y_{1\frac{1}{2}}+y_{-1\frac{1}{2}})+11(y_{2\frac{1}{2}}+y_{-2\frac{1}{2}})\} \qquad \text{......(3)}$$

Fourthly, if

$$y_x = a+bx+cx^2+dx^3$$

$$\int_{-\frac{1}{2}}^{1\frac{1}{2}} y_x\,dx = \tfrac{1}{24}\{27y_0+17y_1+5y_2-y_3\} \qquad \text{......(4)}$$

18. We can now take the calculation of the moments, where $\int_{-\frac{1}{2}}^{n-\frac{1}{2}} y_x\,dx$ is required in terms of $y_0, y_1, \ldots y_{n-1}$.

Now

$$\int_{-\frac{1}{2}}^{n-\frac{1}{2}} y_x\,dx = \int_{-\frac{1}{2}}^{\frac{1}{2}} y_x\,dx+\int_{\frac{1}{2}}^{1\frac{1}{2}} y_x\,dx+\ldots+\int_{n-1\frac{1}{2}}^{n-\frac{1}{2}} y_x\,dx$$

If formula (1) be applied it can be used for all the integrals on the right-hand side of this equation except the first two and the last two, and the values of these are given by (4). Summing the values obtained and writing (4) with the denominator 5760, we obtain

$$\int_{-\frac{1}{2}}^{n-\frac{1}{2}} y_x\,dx = \tfrac{1}{5760}\{6463y_0+4371y_1+6669y_2+5537y_3 +5760(y_4+y_5+\ldots+y_{n-6}+y_{n-5})+5537y_{n-4} +6669y_{n-3}+4371y_{n-2}+6463y_{n-1}\} \qquad \text{......(5)}$$

which means that we can multiply

the first and last ordinates by $\frac{6463}{5760}(=1{\cdot}1220486)$,
the second and last but one by $\frac{4371}{5760}(=0{\cdot}7588542)$,
the third and last but two by $\frac{6669}{5760}(=1{\cdot}1578125)$,
the fourth and last but three by $\frac{5537}{5760}(=0{\cdot}9612847)$,

leave all the other ordinates unaltered, and work out the moments in the usual way from this modified series of ordinates. If there are less than eight ordinates, another formula must be evolved.

19. In the following table the original series and the modified one are set out in the first two columns, and in the other columns the calculations of the first four moments about the middle of the range by the direct method are shown:

TABLE 5

y_x	Modified by formula (5) y'_x	x	$y'_x \times x$	$y'_x \times x^2$	$y'_x \times x^3$	$y'_x \times x^4$
51·81	58·13	−4	232·52	930·08	3720·32	14 881·28
43·74	33·19	−3	99·57	288·71	866·13	2 598·39
35·42	41·01	−2	82·02	164·04	328·08	656·16
27·80	26·72	−1	26·72	26·72	26·72	26·72
20·42	20·42		−440·83	...	−4941·25	...
13·79	13·26	+1	13·26	13·26	13·26	13·26
8·22	9·52	+2	19·04	38·08	76·16	152·32
4·29	3·26	+3	9·78	29·34	88·02	264·06
1·69	1·90	+4	7·60	30·40	121·60	486·40
207·18	207·41		+ 49·68	1520·63	+ 299·04	19 078·59
			−391·15		−4642·21	

207·41 is then treated as the total frequency, and the moments for unit frequency (μ'_n) would be obtained by dividing −391·15, 1520·63, etc. by 207·41, and not by 207·18, which is not the 'total frequency', but merely gives the uncorrected sum of certain equidistant values.

20. The work can sometimes be simplified considerably, for if the values at the ends of the experience are very small and have a tendency to keep close to the axis of x before they finally vanish (i.e. if there is high contact; most actuarial functions l_x, a_x, D_x, etc. have high contact at the old age end of the table), then it is reasonable to suppose that ordinates before the first and after the last exist, but are insignificant in value. Thus the integral corresponding to the whole series of ordinates can be legitimately extended beyond the limits $-\frac{1}{2}$ and $n-\frac{1}{2}$ previously used, because the additional area thus introduced will be evanescent. Now if the area be so extended, the effect will be that in equation (5) the significant ordinates from y_0

to y_{n-1} will all have the coefficient unity, and the ordinates with weighted coefficients will all vanish.

The practical result is, that if there is high contact at one end of the statistics the adjustment need only be made at the other end, while if there is high contact at both ends no adjustment is necessary.

Mathematically, high contact means that all the differential coefficients vanish at the point of contact. The diagrams on pages 63 and 73 show high contact at both ends of the curves, and the diagram on page 56 shows high contact at the longer durations.

21. The second case in § 16, namely, that in which mid-ordinates are used instead of areas, may now be examined. By concentrating areas about the middle points of their bases, we assume that the distances by which the areas $\int_{-\frac{1}{2}}^{\frac{1}{2}} y_x\,dx$, $\int_{\frac{1}{2}}^{1\frac{1}{2}} y_x\,dx$, etc. must be multiplied are the same as the distances from y_0, y_1, etc.; that is, the tth moment from the statistics is

$$\int_{-\frac{1}{2}}^{+\frac{1}{2}} y_x\,dxX^t+\int_{\frac{1}{2}}^{1\frac{1}{2}} y_x\,dx(X+1)^t+\ldots+\int_{n-1\frac{1}{2}}^{n-\frac{1}{2}} y_x\,dx(X+n-1)^t$$

and we require $\int_{-\frac{1}{2}}^{n-\frac{1}{2}} (X+x)^t y_x\,dx$, where X is the distance of y_0 from the ordinate about which moments are calculated.

Applying formula (1) to each integral and collecting terms, we reach as a general coefficient

$$\tfrac{1}{5760}\{\ldots+[5178h^t+308\{(h-1)^t+(h+1)^t\}$$
$$-17\{(h-2)^t+(h+2)^t\}]y+\ldots\}$$

where h is written for $X+x$ for simplification, or

$$\tfrac{1}{5760}\{5760h^t+240t(t-1)h^{t-2}+3t(t-1)(t-2)(t-3)h^{t-4}+\ldots\}$$

If $t=1$ this becomes h

$t=2$ this becomes $h^2+\frac{1}{12}$

$t=3$ this becomes $h^3+\frac{1}{4}h$

$t=4$ this becomes $h^4+\frac{1}{2}h^2+\frac{1}{80}$

It has already been noticed that if there is high contact, the value of $\int_{-\frac{1}{2}}^{n-\frac{1}{2}} (X+x)^t y\,dx$ is found by using the unadjusted ordinates; that is, the second moment is given by a series, the general term of which is h^2y; the third by a series, the general term of which is h^3y, and so on; hence, if μ be written for the true adjusted moment about the mean and ν for the unadjusted moment, the relations between μ and ν are given by

$$\mu_2+\tfrac{1}{12}h^2 = \nu_2 \quad \text{or} \quad \mu_2 = \nu_2-\tfrac{1}{12}h^2$$

$$\mu_4+\tfrac{1}{2}\mu_2h^2+\tfrac{1}{80}h^4 = \nu_4 \quad \text{or} \quad \mu_4 = \nu_4-\tfrac{1}{2}\nu_2h^2+\tfrac{7}{240}h^4$$

The mean needs no adjustment, for if $t = 1$ the general term has the correct coefficient h, and the third moment has to be adjusted by $\frac{1}{4}$ of the first moment, which is zero where the moments are taken about the mean.* In order to demonstrate the correction for the nth moment by the above method, a parabola of at least the nth order is necessary. If we apply these adjustments to the moments found on page 18, for Example 4 of Table 1, we have $\mu_2 = 3{\cdot}1503$, $\mu_3 = -1{\cdot}430976$ and $\mu_4 = 28{\cdot}82866$. These adjustments are found to make an appreciable difference in the constants obtained from the moments, especially when there is a small number of terms.

In other words, they allow for the grouping, and the lesson to be learnt is that a moderate amount of grouping saves work and, thanks to our knowledge of the correct adjustments, we can reduce the average amount of error incurred.

22. The practical conclusions in the two preceding paragraphs as to the treatment of moments when there is high contact can be checked numerically. The equation to a curve with high contact at each end having been written down, we can work out the ordinates at equidistant points or the areas on equal bases and calculate the moments from the figures. From the equation to the curve we can also calculate the area and moments for the whole curve and it will be found that the corresponding figures agree. A good curve with which to make experiments in this way is 'the normal curve of error' because the ordinates and areas are accurately tabulated in *Biometrika Tables*.

A student should calculate the moments for one or two distributions, and make the easier adjustments; he can also find the standard deviations of distributions, for the S.D. $= \sqrt{\mu_2}$, where the μ_2 has been adjusted in accordance with the above rules. In Examples 3 and 4 of Table 1 there is clearly high contact, and in Example 1 the rough moment should be used. In Examples 2 and 5 there is more doubt, and in the calculation of the moments for Example 2 (see p. 54), no adjustment was made.

* These adjustments were first given by W. F. Sheppard in *Proc. Lond. Math. Soc.* XXIX, 353–80. See also Appendix I.

When there is not high contact at both ends of the curve, the adjustments become more difficult to value; suggestions have been made for finding the corrections, and this matter is further discussed in Appendix 1, but a beginner is advised to avoid these refinements.

This advice is given not because adjustment is unnecessary, but because a beginner can content himself with mastering the general idea and leave out some of the refinements until he has a little more experience. Later on, when the methods of Appendix 1 are examined, it will be seen that Sheppard's adjustments alone do not usually improve the rough moments when the distribution is abrupt.

23. Before proceeding to deal with fitting more complicated curves it is advisable to consider the application of the method of moments to a simple case, namely, when

$$y = a+bx+cx^2+\dots$$

Let the range be $2l$, and let the origin be at the middle point of the range, and m_0 stand for the area and m_n for the nth moment of the whole distribution about the middle of the range. Then

$$m_{2s} = \int_{-l}^{+l} (a+bx+cx^2+\dots)x^{2s}\,dx$$

$$= 2l\times l^{2s}\left(\frac{a}{2s+1}+\frac{cl^2}{2s+3}+\dots\right)$$

and similarly $m_{2s+1} = 2l\times l^{2s+1}\left(\dfrac{bl}{2s+3}+\dfrac{dl^3}{2s+5}+\dots\right)$

These equations show that the even moments give the constants a, c, e, etc., and the odd moments give the constants b, d, f, etc. This is, of course, the result of using moments about the middle of the range, and makes the solution of the equations less laborious than they would otherwise have been. The solution can also be simplified a little by writing

$$\frac{1}{2l}\cdot\frac{m_{2s}}{l^{2s}} = \frac{a}{2s+1}+\frac{cl^2}{2s+3}+\dots$$

so that

$$\left.\begin{aligned}\frac{1}{2l}\cdot m_0 &= a+\frac{cl^2}{3}+\frac{el^4}{5}+\dots\\ \frac{1}{2l}\cdot\frac{m_2}{l^2} &= \frac{a}{3}+\frac{cl^2}{5}+\frac{el^4}{7}+\dots\\ \frac{1}{2l}\cdot\frac{m_4}{l^4} &= \frac{a}{5}+\frac{cl^2}{7}+\frac{el^4}{9}+\dots\end{aligned}\right\}$$

and similarly

$$\left.\begin{aligned}\frac{1}{2l}\cdot\frac{m_1}{l}&=\frac{bl}{3}+\frac{dl^3}{5}+\frac{fl^5}{7}+\ldots\\ \frac{1}{2l}\cdot\frac{m_3}{l^3}&=\frac{bl}{5}+\frac{dl^3}{7}+\frac{fl^5}{9}+\ldots\\ \frac{1}{2l}\cdot\frac{m_5}{l^5}&=\frac{bl}{7}+\frac{dl^3}{9}+\frac{fl^5}{11}+\ldots\end{aligned}\right\}$$

The solution of these equations gives the constants required, for example,

(i) if $y = a+bx$, we have

$$a = \frac{1}{2l}m_0$$

$$b = \frac{3}{l}\cdot\frac{1}{2l}\cdot\frac{m_1}{l}$$

(ii) if $y = a+bx+cx^2$

$$a = \frac{3}{4}\left\{\frac{3}{2l}\cdot m_0 - \frac{5}{2l}\cdot\frac{m_2}{l^2}\right\}$$

$$b = \frac{3}{l}\cdot\frac{1}{2l}\cdot\frac{m_1}{l}$$

$$c = \frac{15}{4l^2}\left\{-\frac{1}{2l}\cdot m_0 + \frac{3}{2l}\cdot\frac{m_2}{l^2}\right\}$$

(iii) if $y = a+bx+cx^2+dx^3$

$$a = \frac{3}{4}\left\{\frac{3}{2l}\cdot m_0 - \frac{5}{2l}\cdot\frac{m_2}{l^2}\right\}$$

$$b = \frac{15}{4l}\left\{\frac{5}{2l}\cdot\frac{m_1}{l} - \frac{7}{2l}\cdot\frac{m_3}{l^3}\right\}$$

$$c = \frac{15}{4l^2}\left\{-\frac{1}{2l}\cdot m_0 + \frac{3}{2l}\cdot\frac{m_2}{l^2}\right\}$$

$$d = \frac{35}{4l^3}\left\{-\frac{3}{2l}\cdot\frac{m_1}{l} + \frac{5}{2l}\cdot\frac{m_3}{l^3}\right\}$$

The above results, which can easily be extended if it is wished, may now be applied to one or two numerical examples.

24. As a first example, we shall graduate the statistics in Table 5, § 19, for which the moments about the middle of the range have been calculated. Taking the curve $y = a+bx+cx^2$, the following values from Table 5 will be required:

$$2l = 9 \quad \text{or} \quad l = 4{\cdot}5$$
$$m_0 = 207{\cdot}41$$
$$m_1 = -391{\cdot}15$$
$$m_2 = 1520{\cdot}63$$

Hence
$$a = \frac{3}{4}\left\{\frac{622{\cdot}23}{9} - \frac{5}{9}\times\frac{1520{\cdot}63}{(4{\cdot}5)^2}\right\}$$
$$= 20{\cdot}563$$
$$b = \frac{3}{4{\cdot}5}\times\frac{1}{9}\times\frac{-391{\cdot}15}{4{\cdot}5}$$
$$= -6{\cdot}4387$$
$$c = \frac{15}{4(4{\cdot}5)^2}\left\{-\frac{207{\cdot}41}{9}+\frac{3}{9}\times\frac{1520{\cdot}63}{(4{\cdot}5)^2}\right\}$$
$$= 0{\cdot}36815$$

25. The best way to obtain the ordinates corresponding to this graduation is by calculating $b+c$ the first difference, and $2c$ the second difference, from the middle term; their values are $-6{\cdot}0706$ and $0{\cdot}7363$ respectively. Since second differences are constant, the work is done continuously, and is as follows:

	Δ	Δ^2
52·208	−9·016	0·736
43·192	−8·279	
34·913	−7·543	
27·370	−6·807	
20·563	−6·071	
14·492	−5·335	
9·157	−4·599	
4·558	−3·862	
0·696		

These graduated figures will be found to agree fairly well with those given in the first column of Table 5.

26. As a further example the following statistics, taken from a paper by S. H. J. W. Allin (*J. Inst. Actu.* XXXIX, 350), and giving the values of annuities to widows in pension funds according to the age of the member, may be considered:

Age	Value of annuity	Modified by formula (5) p. 26 a'	Distance from middle of range multiplied by 2 d	$a' \times d$	$a' \times d^2$	$a' \times d^3$
27	21·20	23·79	−7	166·53	1165·71	8159·97
32	19·91	15·11	−5	75·55	377·75	1888·75
37	19·34	22·40	−3	67·20	201·60	604·80
42	18·58	17·86	−1	17·86	17·86	17·86
				−327·14		−10671·38
47	16·74	16·09	+1	16·09	16·09	16·09
52	15·69	18·17	+3	54·51	163·53	490·59
57	14·70	11·15	+5	55·75	278·75	1393·75
62	12·99	14·58	+7	102·06	714·42	5000·94
		139·15		+228·41	2935·71	+6901·37
				−98·73		−3770·01

In calculating the above moments it has been assumed that the figures to be graduated represent a system of ordinates; if they had represented a system of areas, the adjustment by formula (5) would have been unsuitable.

The alternative is to avoid the integral calculus and work out from the equation $y = f(x)$ the sum of the ordinates and the moments of the ordinates. In the particular case where $f(x) = a+bx+cx^2+\ldots$ this is practicable, but there are many expressions which, with their moments, can be integrated but do not lend themselves to finite summation. We have therefore confined attention to the more general method.

When there is an even number of terms the difficulty of calculating the moments about the middle of the range is that the terms have to be multiplied by 0·5, 1·5, 2·5, etc., and if the series to be graduated contains only a few terms, it is best to deal with the distance d, in the way shown above, and then divide the totals by 2, 4 and 8, in order to obtain the first, second and third moments respectively. In this way, we have

$$l = 4$$
$$m_0 = 139{\cdot}15$$
$$m_1 = -\ 49{\cdot}36$$
$$m_2 = 733{\cdot}93$$
$$m_3 = -471{\cdot}25$$

We will now fit the statistics with each of the three curves, the formulae for which have been given, and compare the resulting graduations.

$$\text{(i)}\ y = 17{\cdot}394 - 1{\cdot}157x$$
$$\text{(ii)}\ y = 17{\cdot}633 - 1{\cdot}157x - 0{\cdot}0451x^2$$
$$\text{(iii)}\ y = 17{\cdot}633 - 1{\cdot}190x - 0{\cdot}0451x^2 + 0{\cdot}0035x^3$$

The following table shows the graduations:

Age	Ungraduated	(i)	(ii)	(iii)
27	21·20	21·44	21·13	21·13
32	19·91	20·29	20·24	20·28
37	19·34	19·13	19·27	19·31
42	18·58	17·97	18·20	18·22
47	16·74	16·82	17·04	17·02
52	15·69	15·66	15·80	15·76
57	14·70	14·50	14·46	14·43
62	12·99	13·34	13·03	13·05

Formulae (ii) and (iii) are practically identical, and both are considerably closer to the original figures than (i).

27. The results obtained so far may be summarized as follows:

(1) The method of moments is a general method of finding the constants in a formula suitable to a particular statistical example, and it consists of equating the values of $\sum f(n) \times n^t$ (which is called the tth moment, and is summed for all values of n that occur) to similar expressions obtained from the graduation formula. These latter expressions will be algebraic, and simultaneous equations have to be solved in order to find the arithmetical constants.

(2) The moments from the statistics can be calculated by multiplying the frequencies by appropriate values of n^t or by the summation method.

(3) If moments have been obtained about any one vertical, they can be transferred to any other by the formulae in § 6 of this chapter.

(4) Since the moments from the graduation formula must generally be found by means of the integral calculus, while those from the statistics are found by summation, the latter have to be adjusted before the equations for obtaining the constants can be correctly formed. The adjustments depend on whether the statistics are a system of ordinates or a system of areas; in the former case adjustment is made by equation (5), and in the latter by the formulae in § 21 if there is high contact at both ends of the curve.

4. Pearson's system of frequency-curves

1. When it becomes necessary in practical work to decide on a system of curves for describing frequency distributions, we have to bear in mind that

(1) Any expression used must be a graduation formula; it must remove the roughness of the material.

(2) The formula must not be too complicated. In particular there must not be too many constants in the formula. Also we wish to avoid use of high moments. The higher the moment the more liable it is to error when deduced from ungraduated observations; this is clear, when we remember that the ends of the experiences are multiplied by the highest numbers and their powers.

(3) There must be a systematic method of approaching frequency distributions.

2. Now, considering the more obvious characteristics of frequency distributions, we find they generally start at zero, rise to a maximum, and then fall sometimes at the same but often at a different rate. At the ends of the distribution there is often high contact. This means, mathematically, that a series of equations $y = f(x)$, $y = \phi(x)$, etc., must be chosen, so that in each equation of the series $dy/dx = 0$ for certain values of x, namely, at the maximum and at the end of the curve where there is contact with the axis of x.

The above suggests that dy/dx may be put equal to $[y \times (x+a)]/F(x)$; then, if $y = 0$, $dy/dx = 0$, and there is, therefore, contact at one end, of the curve, while if $x = -a$, $dy/dx = 0$, and we have the maximum we require. So long as $F(x)$ is general the form assumed for dy/dx is extremely general and includes cases when dy/dx may not be zero when y is zero. If $F(x)$ is expanded by Maclaurin's theorem in ascending powers of x, we have

$$\frac{dy}{dx} = \frac{y(x+a)}{b_0 + b_1 x + b_2 x^2 + \dots} \qquad \text{......(1)}$$

We shall return to this equation and show how it can be put in the form $y = f(x)$, so as to express y as a direct function of x, and we shall see that we have obtained something more general than is implied at the beginning of this paragraph. We shall obtain curves taking various widely different shapes. As the matter has up to the present been approached from an experimental point of view, it will be interesting to see how equation (1) can be obtained up to the x^2 term in the denominator from elementary propositions in the theory of probabilities.

3. If p be the probability of an event happening and q the probability of its failing, then the probabilities of its happening once, twice, and so on out of n trials are given by the terms of the expansion $(p+q)^n$; or if we have N cases, the terms of $N(p+q)^n$ give the the frequency distribution of the N cases into $n+1$ groups. The binomial series does not represent nearly all the probabilities that arise, and another series that occurs is the hypergeometrical. Thus the chances of getting $r, r-1, \ldots, 0$ black balls from a bag containing pn black and qn white balls when r balls are drawn, are given by the successive terms of the series

$$\frac{pn(pn-1)\ldots(pn-r+1)}{n(n-1)\ldots(n-r+1)} \times\left\{1+\frac{rqn}{pn-r+1}+\frac{r(r-1)}{2!}\frac{qn(qn-1)}{(pn-r+1)(pn-r+2)}+\ldots\right\}$$

Other series may arise, but those given will be sufficient for the present purpose, and we shall proceed to consider how they can be put in the form of equation (1). The inconvenience of the expressions as they now stand becomes fairly obvious when an attempt is made to calculate numerical values for a large number of groups, and besides this, they are not continuous, while the statistics of practical work often are.

Considering the hypergeometrical series, and remembering that the function required for equation (1) is $(1/y)(dy/dx)$, and that, as the series is discontinuous, finite differences must be used, we have

$$y_x=\frac{pn(pn-1)\ldots(pn-r+1)}{n(n-1)\ldots(n-r+1)}\cdot\frac{r(r-1)\ldots(r-x+2)}{(x-1)!} \times\frac{qn(qn-1)\ldots(qn-x+2)}{(pn-r+1)(pn-r+2)\ldots(pn-r+x-1)}$$

$$\Delta y_x = y_{x+1}-y_x = y_x\left\{\frac{r-x+1}{x}\cdot\frac{qn-x+1}{pn-r+x}+1\right\}$$

$$= y_x\left\{\frac{(r+1)(qn+1)-x(n+2)}{x(pn-r+x)}\right\} \quad \text{for} \quad p+q=1$$

and

$$y_{x+\frac{1}{2}} = \tfrac{1}{2}(y_{x+1}+y_x)$$
$$= \tfrac{1}{2}y_x\left\{\frac{(r+1)(qn+1)-x[2(r+2)+n(q-p)]+2x^2}{x(pn-r+x)}\right\}$$

$$\therefore \qquad \frac{\Delta y_x}{y_{x+\frac{1}{2}}} = \frac{2\{(r+1)(qn+1)-x(n+2)\}}{(r+1)(qn+1)-x\{2(r+1)+n(q-p)\}+2x^2}$$

which may be put in the form of equation (1),

$$\frac{1}{y}\frac{dy}{dx} = \frac{a+x}{b_0+b_1x+b_2x^2} \qquad \text{......(1)}'$$

In this form the actuarial reader will naturally think of the force of mortality: to proceed from the force of mortality, after changing its sign, to the 'number living' (l_x) in a life table is the same thing as to proceed from the formula just given to a frequency-curve.

4. Returning to equation (1), we see that it can be written in the form

$$(b_0+b_1x+b_2x^2+\ldots)\frac{dy}{dx} = y(x+a)$$

Multiplying each side by x^n, and integrating with respect to x, we have

$$\int x^n(b_0+b_1x+b_2x^2+\ldots)\frac{dy}{dx}\,dx = \int y(x+a)x^n dx$$

Integrate the left-hand side by parts treating dy/dx as one part, and the right-hand side as the sum of two functions, and then

$$x^n(b_0+b_1x+b_2x^2+\ldots)y-\int \{nb_0x^{n-1}+(n+1)b_1x^n+$$
$$(n+2)b_2x^{n+1}+\ldots\}ydx$$
$$= \int yx^{n-1}dx+\int yax^n dx$$

or, if at the ends of the range of the curve the expression

$$x^n(b_0+b_1x+b_2x^2+\ldots)y$$

vanishes, we have

$$-nb_0\mu'_{n-1}-(n+1)b_1\mu'_n-(n+2)b_2\mu'_{n+1}-\ldots = \mu'_{n+1}+a\mu'_n$$

where we use the notation we have already adopted, namely,

$$\mu'_n = \int yx^n\,dx$$

If we put $n = 0, 1, 2, \ldots s$ respectively, we get $s+1$ equations to enable us to find $a, b_0, b_1, \ldots$ etc., in terms of the moments (μ') as shown by the following equations, which have been obtained by writing the equation in the form

$$a\mu'_n+nb_0\mu'_{n-1}+(n+1)b_1\mu'_n+(n+2)b_2\mu'_{n+1}+\ldots = -\mu'_{n+1}$$

and then putting $n = 0, 1, 2$, etc.

$$\left.\begin{aligned} a\mu'_0+0\times b_0+\ b_1\mu'_0+2b_2\mu'_1+\ldots &= -\mu'_1 \\ a\mu'_1+\ b_0\mu'_0+2b_1\mu'_1+3b_2\mu'_2+\ldots &= -\mu'_2 \\ a\mu'_2+2b_0\mu'_1+3b_1\mu'_2+4b_2\mu'_3+\ldots &= -\mu'_3 \\ a\mu'_3+3b_0\mu'_2+4b_1\mu'_3+5b_2\mu'_4+\ldots &= -\mu'_4 \end{aligned}\right\} \quad \ldots\ldots(2)$$

etc.

Let us now make $\mu'_1 = 0$, and alter the other moments in the way indicated in Chapter 3, for the result of making $\mu'_1 = 0$ is to change the origin of the system to the mean of the distribution. We can also treat μ'_0 as 1, and these simplifications lead to the following results:

(1) Keeping b_0 only, we have

$$\frac{1}{y}\frac{dy}{dx} = -\frac{x}{\mu_2}$$

(2) Keeping b_0 and b_1, the first three equations in the system (2) above give

$$a+b_1 = 0$$

$$b_0 = -\mu_2$$

and $$a\mu_2+3b_1\mu_2 = -\mu_3$$

or $$b_1 = -\frac{\mu_3}{2\mu_2}$$

and $$a = \frac{\mu_3}{2\mu_2}$$

and the differential equation becomes

$$\frac{1}{y}\frac{dy}{dx} = -\frac{\mu_2+\dfrac{\mu_3}{2\mu_2}x}{x+\dfrac{\mu_3}{2\mu_2}}$$

(3) Keeping b_0, b_1, b_2, the system gives

$$a+b_1 = 0$$

$$b_0+3b_2\mu_2 = -\mu_2$$

$$a\mu_2+3b_1\mu_2+4b_2\mu_3 = -\mu_3$$

$$a\mu_3+3b_0\mu_2+4b_1\mu_3+5b_2\mu_4 = -\mu_4$$

The solution of these simultaneous equations is perfectly straightforward, and leads to

$$\frac{1}{y}\frac{dy}{dx}=-\frac{x+\dfrac{\mu_3(\mu_4+3\mu_2^2)}{10\mu_2\mu_4-18\mu_2^3-12\mu_3^2}}{\dfrac{\mu_2(4\mu_2\mu_4-3\mu_3^2)}{10\mu_2\mu_4-18\mu_2^3-12\mu_3^2}+\dfrac{\mu_3(\mu_4+3\mu_2^2)}{10\mu_2\mu_4-18\mu_2^3-12\mu_3^2}x+\dfrac{2\mu_2\mu_4-3\mu_3^2-6\mu_2^3}{10\mu_2\mu_4-18\mu_2^3\ \ 12\mu_3^2}x^2}$$

In this last form put $\sqrt{\beta_1}=\mu_3/\mu_2^{\frac{3}{2}}$ and $\beta_2=\mu_4/\mu_2^2$ and

$$\frac{1}{y}\frac{dy}{dx}=-\frac{x+\dfrac{\sqrt{\mu_2}\sqrt{\beta_1}(\beta_2+3)}{2(5\beta_2-6\beta_1-9)}}{\dfrac{\mu_2(4\beta_2-3\beta_1)+\sqrt{\mu_2}\sqrt{\beta_1}(\beta_2+3)x+(2\beta_2-3\beta_1-6)x^2}{2(5\beta_2-6\beta_1-9)}} \quad \ldots\ldots(3)$$

Note that b_0 cannot be zero, since $4\beta_2>3\beta_1$. (In fact $\beta_2-\beta_1-1\geqslant 0$, so $4\beta_2-3\beta_1\geqslant 4\beta_2-4\beta_1-4\geqslant 0$.)

A similar argument * leads to the conclusion that the mean deviation is equal to

$$\tfrac{1}{3}\mu_2(y_0/N)(4\beta_2-3\beta_1)/(\beta_2-\beta_1-1)$$

5. The reasoning by which equation (1) was first obtained showed that a is the distance between the origin and the mode, or as the origin has now been transferred to the mean by putting $\mu_1=0$, a is the distance between the mean and the mode. This distance in terms of the moments is, therefore,

$$\frac{\sigma\sqrt{\beta_1}(\beta_2+3)}{2(5\beta_2-6\beta_1-9)}$$

where σ is the standard deviation $\sqrt{\mu_2}$.

Since the skewness is the distance between the mean and mode divided by the standard deviation

$$\text{Skewness}=\frac{\sqrt{\beta_1}(\beta_2+3)}{2(5\beta_2-6\beta_1-9)}$$

6. It would be possible to obtain constants in the differential equation (1) by using a greater number of terms and retaining b_3, b_4, etc., but there are strong practical objections to such a course. Besides the increase in arithmetical work, the gain in introducing additional constants is small because the higher moments become untrustworthy, as we have already noticed. Karl Pearson has shown † that 'we might easily on a random sample reach a 7th or 8th

* K. Pearson, *Biometrika*, XVI (1924), 196.

† 'Skew correlation and non-linear regression', *Drapers' Company Res. Mem.* 1905, p. 9. See also Chapter 10.

moment having half or double the value it actually has in the general population. Constants based on these high moments will be practically idle. They may enable us to describe closely an individual random sample, but no safe argument can be drawn from this individual sample as to the general population at large, at any rate so far as the argument is based on the constants depending on these high moments.' In some actuarial statistics where there are as many as 100 000 cases, it might be worth while to go as far as the next term of the series, but even here the value of the work is discounted because any other smaller body of statistics on the same subject could not be compared satisfactorily with the result. For practical purposes it is probable that the equation taken as far as b_2 will be sufficient, and we shall confine our attention to the forms thus obtained.

7. Turning to the particular form of equation (1) given in equation (3) it will be seen that it is possible to obtain a formula representing the statistics by inserting in that equation the values of the moments found from the statistics, but this would not give a graduation in the same form as that in which the original data appeared, for in the latter we have y, while the former gives $(1/y)(dy/dx)$ or $(d \log y)/dx$. It would, therefore, be necessary to integrate the expression we obtain in order to get terms comparable with the original data, and it is better in practical work to deal with the equations in the forms in which we require them for comparison, rather than by using the differential equation and then integrating the result.

8. The next step is, therefore, to replace the equation

$$\frac{d \log y}{dx} = \frac{x+a}{b_0+b_1x+b_2x^2}$$

by one of the form $y = f(x)$, and to do this $(x+a)/(b_0+b_1x+b_2x^2)$ must be integrated.

Let us consider equation (3) as a general expression for integration, then we notice that the form the integral takes depends on the particular values of the coefficients of x in the denominator. The problem is, in fact, merely a consideration of the forms taken by the denominator for

$$b_1x+b_2x^2$$

$$= b_2\left[x - \frac{-b_1+\sqrt{(b_1^2-4b_0b_2)}}{2b_2}\right]\left[x - \frac{-b_1-\sqrt{(b_1^2-4b_0b_2)}}{2b_2}\right]$$

and the criterion for fixing the form in a particular case is, obviously, the same as that for the nature of the roots of the equation $b_0+b_1x+b_2x^2=0$, viz. $b_1^2/(4b_0b_2)$, which, by substituting from formula (3), gives

$$\frac{\beta_1(\beta_2+3)^2}{4(2\beta_2-3\beta_1-6)(4\beta_2-3\beta_1)} \qquad \text{......(4)}$$

We shall denote this by κ, and refer to it as 'the criterion'.

9. If expression (4) is negative the roots are real and of different sign, and we get one of the main types of curve—called Type I by Karl Pearson, to whom this system of curves is due; if expression (4) is positive and less than unity the roots are complex, and we get the second main type (Pearson's Type IV), and if expression (4) is positive and greater than unity the roots are real and of the same sign, and we reach the third main type (Pearson's Type VI).

This really covers the whole field, but in the limiting cases when one type changes into another we reach simpler forms of transition curves. Thus when the criterion is large (theoretically infinite) one root is ∞ (Type III), when it is unity the two roots are equal (Type V), and when it is zero the roots are equal in magnitude but of opposite sign (Type II). If in the last case $b_1 = b_2 = 0$, we reach what we shall call the 'normal curve of error': this name is open to some objection just as are the other names given to it (e.g. Probability curve, Gaussian curve, etc.). Then again the expression for $(d \log y)/dx$ may be reducible to the form $a'/(b_0'+b_1'x)$ and perhaps a straight line for the frequency-curve (cf. Types VIII, IX and XI), while if the expression reduces to a constant the curve is the ordinary geometrical progression which we are pleased to find as a special case of a system of frequency-curves because we are already familiar with it in the theory of probability in connection with sequences from coin tossing, etc. As we proceed we shall find that in certain circumstances the curves may be **J**-shaped or even **U**-shaped, with limits of a single ordinate or two separated ordinates. A diagram at the end of the book will give the reader an idea of the variety of shapes taken by the curves evolved from the formula

$$\frac{d \log y}{dx} = \frac{x+a}{b_0+b_1x+b_2x^2}$$

In practice we shall require the equations to the various kinds of frequency-curve, and we shall also want to know which type should be used in a particular case. We cannot usually guess the type from the appearance of the rough data and need an arithmetical test.

10. We will deal first with the equations to the frequency-curves, that is, with the actual integration, and begin with the three main types.

First Main Type (*Pearson's Type I*). The factors in the denominator, when the roots of $b_0+b_1x+b_2x^2=0$ are real and of different signs, take the form

$$b_2\left[x-\frac{-b_1+\sqrt{\text{a positive quantity}}}{2b_2}\right]\times\left[x-\frac{-b_1-\sqrt{\text{a positive quantity}}}{2b_2}\right]$$

and the expression to be integrated is therefore of the form

$$\frac{1}{b_2}\cdot\frac{x+a}{(x+A_1)(x-A_2)}=\frac{1}{b_2}\cdot\frac{A_1-a}{A_1+A_2}\cdot\frac{1}{x+A_1}-\frac{1}{b_2}\cdot\frac{A_2+a}{A_1+A_2}\cdot\frac{1}{A_2-x}$$

by partial fractions (with $A_1, A_2>0$).

The integration is now simple, and gives

$$\log y=\frac{1}{b_2}\cdot\frac{A_1-a}{A_1+A_2}\log(x+A_1)+\frac{1}{b_2}\cdot\frac{A_2+a}{A_1+A_2}\log(A_2-x)+\text{a constant}$$

$$\therefore\qquad y=y'(x+A_1)^{\frac{1}{b_2}\cdot\frac{A_1-a}{A_1+A_2}}(A_2-x)^{\frac{1}{b_2}\cdot\frac{A_2+a}{A_1+A_2}}$$

where y' results from the constant introduced by integration.

Since y cannot be negative over the range of variation, this should be $-A_1<x<A_2$.

If the origin is now transferred to the mode (i.e. put x for $x+a$), we have

$$y=y_0\left(1+\frac{x}{a_1}\right)^{m_1}\left(1-\frac{x}{a_2}\right)^{m_2}$$

where $a_1=A_1-a$, $a_2=A_2+a$, and $m_1/a_1=m_2/a_2$.

Second Main Type (*Pearson's Type IV*). If the roots of the equation $b_0+b_1x+b_2x^2=0$ are complex, it is impossible to throw the denominator into real factors; and when this occurs we have to integrate by putting the expression on the right-hand side of the fundamental differential equation in the form

$$\frac{X+c}{b_2(X^2+A^2)}$$

where $X=x+\dfrac{b_1}{2b_2}$, $\quad c=a-\dfrac{b_1}{2b_2}\quad$ and $\quad A^2=\dfrac{b_0}{b_2}-\dfrac{b_1^2}{4b_2^2}$

Then
$$\log y = \int \frac{X+c}{b_2(X^2+A^2)}\,dX$$
$$= \int \frac{X}{b_2(X^2+A^2)}\,dX + \int \frac{c}{b_2(X^2+A^2)}\,dX$$
$$= \frac{1}{2b_2}\log(X^2+A^2) + \frac{c}{Ab_2}\tan^{-1}\frac{X}{A} + \text{constant}$$
$\therefore$
$$y = y'(X^2+A^2)^{1/2b_2}e^{(c/Ab_2)\tan^{-1}X/A}$$
or
$$y = \frac{y_0}{\left(1+\frac{x^2}{a^2}\right)^m}e^{-\nu\tan^{-1}x/a}$$

where $m = -1/(2b_2)$; $\nu = -c/(Ab_2)$; $a = A$. Note that a has a meaning different from that implied in equation (2). The relation between this type and Type I can be seen by factorizing the denominator of the right-hand side of the differential equation, $b_2(X-iA)(X+iA)$, and then obtaining an expression for y having the same form as Type I, but containing complex expressions.

Third Main Type (*Pearson's Type VI*). The factorizing is the same as Type I, but the roots of the equation being of like sign, the factors of the denominator take the form $(x+A_1)(x+A_2)$. The work is then the same, but at the end the origin is put by Pearson not at the mode but so that one of the expressions $x+A_1$ or $x+A_2$ can be written as x. The form is then
$$y = y_0(x-a)^{m_1}x^{-m_2}$$

11. We may now set out a few of the transition types.

Pearson's *Type II* is the same as his Type I with $a_1 = a_2$ and so $m_1 = m_2$.

Type III. This type is reached when the criterion is ∞, which happens when $b_2 = 0$. (As pointed out above, b_0 cannot be zero.)
$$\log y = \int \frac{x+a}{b_0+b_1x}\,dx$$
$$= \int\left(\frac{1}{b_1} + \frac{a-b_0/b_1}{b_1x+b_0}\right)dx$$
$$= \frac{x}{b_1} + \left(a-\frac{b_0}{b_1}\right)\frac{1}{b_1}\log(b_1x+b_0) + \text{constant}$$
and
$$y = y'e^{x/b_1}(b_1x+b_0)^{(a-b_0/b_1)/b_1}$$
or, by changing the origin,
$$y = y_0e^{-\gamma x}\left(1+\frac{x}{a}\right)^{\gamma a} \quad \text{for} \quad x > -a.$$

where a has a meaning different from that implied in equation (1). This type can be seen to be a particular case of Type I when a_2 becomes infinite.

Type V. In this case, when the roots are real and equal,

$$\log y = \int \frac{1}{b_2} \frac{x+a}{(x+b_1/2b_2)^2} dx$$

$$= \int \frac{1}{b_2} \frac{(x+b_1/2b_2)+(a-b_1/2b_2)}{(x+b_1/2b_2)^2} dx$$

$$= \int \frac{dx}{b_2(x+b_1/2b_2)} + \int \frac{a-b_1/2b_2}{b_2(x+b_1/2b_2)^2} dx$$

$$= \frac{1}{b_2} \log(x+b_1/2b_2) - \frac{a-b_1/2b_2}{b_2(x+b_1/2b_2)} + \text{constant}$$

$$\therefore \qquad y = y'(x+b_1/2b_2)^{\frac{1}{b_2}} e^{-\frac{a-b_1/2b_2}{b_2(x+b_1/2b_2)}} \text{ or, moving the origin to } -b_1/(2b_2)$$

$$y = y_0 x^{-p} e^{-\gamma/x} \qquad (x>0)$$

Normal Curve of Error. Putting

$$b_1 = b_2 = 0$$

$$\log y = \int \frac{x+a}{b_0} dx$$

$$= \frac{x^2}{2b_0} + \frac{ax}{b_0} + \text{constant}$$

$$= \frac{(x+a)^2}{2b_0} + \text{constant}$$

$$\therefore \qquad y = y' e^{(x+a)^2/2b_0}$$

or, by changing the origin and altering the constant,

$$y = y_0 e^{-x^2/c}$$

In a similar way the other less important transition curves can be obtained. These are

$$\left(1+\frac{x}{a}\right)^{-m}; \qquad \left(1+\frac{x}{a}\right)^{m}; \qquad e^{-x/\sigma}; \qquad x^{-m}$$

and we reach **J**-shaped curves when in Type I either m_1 or m_2 is negative and **U**-shaped curves when both are negative. **J**-shaped curves can also be obtained with Type III.

12. A table is inserted which gives the list of curves with Pearson's numbering and with the origin as he generally uses it. This is convenient because in reading other work on the subject it will be found

TABLE 6. *Frequency-curves*

No. of type usually adopted (Pearson)	Equation to curve in form usually adopted (Pearson): Equation	Origin	Equation with origin at mean	Criterion	Remarks	For calculation of constants see page
MAIN TYPES						
I	$y = y_0(1+x/a_1)^{\nu a_1}(1-x/a_2)^{\nu a_2}$	Mode (antimode)	$y = y_e(1+x/A_1)^{m_1}(1-x/A_2)^{m_2}$ where $(m_1+1)/A_1 = (m_2+1)/A_2$	κ negative	Limited range ($-a_1$ to a_2); skew; usually bell-shaped, but may be **U**-shaped, **J**-shaped or twisted **J**-shaped	51–2
IV	$y = y_0(1+x^2/a^2)^{-m}e^{-\nu\tan^{-1}x/a}$	$\nu a/(2m-2)$ after mean	$y = y_0[1+(x/a-\nu/r)^2]^{-m}e^{-\nu\tan^{-1}(x/a-\nu/r)}$ where $r = 2m-2$	$\kappa>0$ and <1	Unlimited range; skew; bell-shaped	58–9
VI	$y = y_0(x-a)^{q_2}x^{-q_1}$	a before start of curve	$y = y_e(1+x/A_1)^{-q_1}(1+x/A_2)^{q_2}$ where $(q_1-1)/A_1 = (q_2+1)/A_2$	$\kappa>1$	Unlimited range in one direction (a to ∞); skew; bell-shaped, but may be **J**-shaped	67
TRANSITION TYPES						
'Normal' curve'	$y = y_0e^{-x^2/2\sigma^2}$	Mode (= mean)	$y = y_0e^{-x^2/2\sigma^2}$	$\kappa = 0,\ \beta_1 = 0,\ \beta_2 = 3$	Unlimited range; symmetrical; bell-shaped	70
II	$y = y_0(1-x^2/a^2)^m$	Mode (= mean)	$y = y_0(1-x^2/a^2)^m$	$\kappa = 0,\ \beta_1 = 0,\ \beta_2<3$	Limited range ($-a$ to a); symmetrical; usually bell-shaped, but **U**-shaped when $\beta_2<1{\cdot}8$	74
VII	$y = y_0(1+x^2/a^2)^{-m}$	Mode (= mean)	$y = y_0(1+x^2/a^2)^{-m}$	$\kappa = 0,\ \beta_1 = 0,\ \beta_2>3$	Unlimited range; symmetrical; bell-shaped	77
III	$y = y_0(1+x/a)^{\gamma a}e^{-\gamma x}$	Mode	$y = y_e(1+x/A)^pe^{-\gamma x}$ where $A = (p+1)/\gamma$ and $p = \gamma a$	$2\beta_2 = 6+3\beta_1$	Unlimited range in one direction ($-a$ to ∞); usually bell-shaped, but may be **J**-shaped	78–9
V	$y = y_0x^{-p}e^{-\gamma/x}$	Start of curve	$y = y_e(1+x/A)^{-p}e^{(p-2)/(1+x/A)}$	$\kappa = 1$	Unlimited range in one direction (0 to ∞); bell-shaped	81–2
VIII	$y = y_0(1+x/a)^{-m}$	End of curve	$y = y_e(1+x/A)^{-m}$	κ negative; $\lambda = 0$; $5\beta_2-6\beta_1-9$ negative	Range from infinite ordinate at $-a$ to finite ordinate at 0 (or from $-a(1-m)/(2-m)$ to $a/(2-m)$ with origin at mean)	86
IX	$y = y_0(1+x/a)^m$	End of curve	$y = y_e(1+x/A)^m$	κ negative; $\lambda = 0$; $5\beta_2-6\beta_1-9$ positive, $2\beta_2-3\beta_1-6$ negative	Range from $x = -a$ where $y = 0$ to $x = 0$ where $y = y_0$ (or from $-a(m+1)/(m+2)$ to $a/(m+2)$ with origin at mean)	89
X	$y = y_0e^{-x/\sigma}$	Start of curve	$y = y_ee^{-x/\sigma}$	$\beta_1 = 4,\ \beta_2 = 9$	Exponential from finite ordinate at 0 (or $-\sigma$ with origin at mean) to infinitesimal ordinate at ∞; **J**-shaped	91
XI	$y = y_0x^{-m}$	b before start	$y = y_e(1+x/A)^{-m}$	$\kappa>1,\ \lambda = 0$, $2\beta_2-3\beta_1-6$ positive	**J**-shaped; starts at $x = b$ (or $-b/(m-2)$ with origin at mean) where ordinate is finite	92–3
XII	$y = y_0\left(\dfrac{\sigma\{\sqrt{(3+\beta_1)}+\sqrt{\beta_1}\}+x}{\sigma\{\sqrt{(3+\beta_1)}-\sqrt{\beta_1}\}-x}\right)^{\sqrt{(\beta_1/(3+\beta_1))}}$	Mean	$y = y_e(1+x/A_1)^{m_1}(1-x/A_2)^{m_2}$ is an alternative form where m_1 and m_2 are equal numerically and <1 but of opposite sign	$5\beta_2-6\beta_1-9 = 0$	Twisted **J**-shaped; special case of Type I	94

$$\kappa = \frac{\beta_1(\beta_2+3)^2}{4(4\beta_2-3\beta_1)(2\beta_2-3\beta_1-6)},\qquad \beta_1 = \mu_3^2/\mu_2^3,\qquad \beta_2 = \mu_4/\mu_2^2,\qquad \lambda = \frac{(4\beta_2-3\beta_1)(10\beta_2-12\beta_1-18)^2-\beta_1(\beta_2+3)^2(8\beta_2-9\beta_1-12)}{(3\beta_1-2\beta_2+6)\{\beta_1(\beta_2+3)^2+4(4\beta_2-3\beta_1)(3\beta\ -2\beta_2+6)}$$

that Pearson's numbering, etc., is usually adopted. I have, however, added a note of the equation to each curve when the origin is at the mean. A column in the table gives criteria to show which curve should be used in an individual case.

We may here deal with a little difficulty that students sometimes encounter in connection with types which may be expressed in the same algebraic form (e.g. Types VIII, IX and XI can all be written hx^k). The question may be asked why we should not fit hx^k from a to b and find h, k, a and b from the equations for the moments. The answer is that the criteria afford in effect a simplification of the equations and automatically tell us a good deal about the value of the constants and the range of the curve.

13. We shall return to some of the technical points when discussing numerical examples in the next chapter but may now recapitulate the method, and see the steps that have to be taken to fit a frequency curve to statistics.

(1) Arrange the statistics in sequence.

(2) Calculate the moments about a convenient vertical.

(3) Transfer the moments to the centroid vertical (vertical through the mean).

(4) If there is high contact at both ends of the curve, apply Sheppard's adjustments to the moments (i.e. deduct $\frac{1}{12}$ and $\frac{1}{2}\nu_2 - \frac{7}{240}$ from the second and fourth moments respectively). If there is not high contact, see Appendix 1.

(5) Calculate the criterion.

(6) By means of Table 6 decide which curve should be used.

As an alternative to (5) and (6), the curve to be used can be found from diagrams in *Biometrika Tables*, which show the type in terms of β_1 and β_2.

5. Calculation

1. The next point to be considered is the calculation of the constants for any particular distribution, when the moments have been calculated and the type to be used has been decided. The formulae required for the numerical work will be given for each type; a numerical example, including the calculation of the graduated figures, will follow, with the proofs of the formulae.

2. Some general points relating to the calculation of the curves when the constants have been found may be conveniently considered here. When the constants are known, we can calculate the ordinate for any value of x by substituting that value in the expression for the frequency-curve; and if areas are required, some method of proceeding from ordinates to areas must be found. The most simple is probably to calculate mid-ordinates, and then by the quadrature formula (1) or (2) on page 26 find the areas. It is occasionally more convenient to calculate the ordinates at the beginning of each group, and then formula (3) should be used. These formulae can be best applied in the form of differences; thus, from (2) we have

$$\int_{-\frac{1}{2}}^{\frac{1}{2}} y_x\,dx = y_0 - \tfrac{1}{24}\{\Delta y_{-1} - \Delta y_0\}$$

from (1)

$$\int_{-\frac{1}{2}}^{\frac{1}{2}} y_x\,dx = y_0 - \tfrac{291}{5760}\{\Delta y_{-1} - \Delta y_0\} + \tfrac{17}{5760}\{\Delta y_{-2} - \Delta y_1\}$$

from (3)

$$\int_{-\frac{1}{2}}^{\frac{1}{2}} y_x\,dx = \tfrac{1}{2}\{y_{\frac{1}{2}} + y_{-\frac{1}{2}}\} + \tfrac{82}{1440}\{\Delta y_{-1\frac{1}{2}} - \Delta y_{\frac{1}{2}}\} - \tfrac{11}{1440}\{\Delta y_{-2\frac{1}{2}} - \Delta y_{1\frac{1}{2}}\}$$

Formula (2) is generally sufficiently accurate, while the others will be found to give a result true to five figures in ordinary cases—exceptional cases will be referred to in the numerical examples that follow.

Another convenient formula is Simpson's rule

$$\int_{-\frac{1}{2}}^{\frac{1}{2}} y_x\,dx = \tfrac{1}{6}(y_{-\frac{1}{2}} + 4y_0 + y_{\frac{1}{2}})$$

In terms of central differences (with interval $\frac{1}{2}$), this is $y_0+\frac{1}{6}\delta^2 y_0$. If y_{-1} and y_1 are also used, a further term $-\frac{1}{180}\delta^4 y_0$ is added. This term is usually small.

3. It is sometimes a help to see the graduation expressed graphically, and this has been done with some of the examples. The best method is to insert the ordinate at the mode; note the ends of the curve, and the heights of the ordinates that have been calculated. These heights give points on the curve, which can be drawn through them fairly easily. In drawing the curve, as well as in calculating the constants, the sign of the skewness must be borne in mind, for it is possible to draw the curve with the skewness on the wrong side of the mode, and if the distribution is nearly symmetrical, it is not so easy to notice the mistake as one might expect. The tangent to the curve at the mode is parallel to the axis of x except in the case of the **J**-shaped curves or some of the less common transition types.

4. It is best to draw on a rather large scale in order to gain distinctness, and the curves given here were drawn larger than their present size; the reduction being, of course, made in the process of reproduction.

The base elements should also be fairly large in proportion to the height, so that the curve may not ascend too steeply; otherwise small horizontal differences between the graduated and ungraduated curves are apt to conceal large vertical differences when the curve is rising or falling rapidly, but it is the latter differences that are of importance.

5. The reader should notice that all the cases considered in the following pages assume complete distributions, and it is in general only possible to find the curve from part of a distribution by means of successive approximation which is extremely laborious. Another point, to which reference will again be made, is with regard to grouping statistics; it is sometimes impossible to obtain many groups, but for accuracy in finding moments the greater the number of groups the better, unless the total number of cases is small. A little discretion is needed in this respect, but in actuarial statistics which are sometimes based on as many as 200 000 cases, seventy groups might be used for great accuracy. In our examples we have grouped merely to save work, space and printing, and the grouping does not alter the method.

If there is high contact so that we know the proper adjustments, grouping leads to little or no error. An adjustment of one-twelfth to the second moment when ten ages are grouped and used as the unit has much more effect, proportionately, than when only five ages are grouped or when individual ages are used. The fear sometimes expressed that grouping destroys accuracy has no proper foundation in such cases; a little numerical evidence on this point will be found in Appendix 1.

6. Another matter with which it seems advisable to deal here is connected with the criterion, κ. This may have any value from $-\infty$ to $+\infty$, and from the following diagram it will be seen how the types cover all the possible values of the criterion and do not overlap.

$\kappa = -\infty$		$\kappa = 0$		$\kappa = 1$		$\kappa = \infty$
	κ negative		$\kappa>0$ and <1		$\kappa>1$	
	Type I		Type IV		Type VI	
Type III		Normal curve when $\beta_2 = 3$ Type II (VII) when $\beta_2 < 3$ (>3)		Type V		Type III

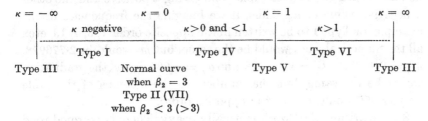

Just before $\kappa = 0$, Type I becomes nearly symmetrical, and after that value is passed we have a skew curve of unlimited range, and so on. At each critical point there are one or more 'transition' curves. If by a mistake a student should use the wrong main type, he will find his mistake by reaching an imaginary quantity in one of the square roots which occur in the equations for the constants, but transition types can be used when the values of the criterion approximate to the theoretical values; they can, in fact, be viewed as approximations which give an accurate result in a limiting case. It is impossible to give exact limits within which we are justified in using a transition type; theoretically the justification depends on the size of the standard error of the function dealt with, but in practice we can be guided to a great extent by the size of the experience; if there are few cases, a larger deviation in the criterion will arise than if there are many. Individual cases must be considered on their merits, but if the student finds himself in doubt he can avoid using the transition type and be on the safe side in the matter of accuracy. The student has one safe guide in every case, namely, that 'the proof of the pudding is in the eating'. He should try transition curves in a few cases where he has little hope of their applicability and compare the results with those obtained by the right main types and he will then learn much about both classes of curves.

7. In the formulae that are given for the various types, the choice of sign for a square root depends on the sign of μ_3. If the frequency is concentrated more closely below the mean than above it, the mode is on the left-hand side of the mean and μ_3 is positive; the signs of certain constants in each type must therefore depend on the sign of μ_3 in order that the mode and mean may lie in their correct relative positions.* Where, however, no remark is made as to the sign of the expression in which a square root is given the positive root is implied, and the reader will find that these rules become easier to follow when he has worked out two examples, one giving a positive and the other a negative value of μ_3. Thus, if we imagine the frequencies in the example for Type I to be written in the opposite order 1, 3, 7, 13, etc., all the numerical work would be the same, but m_1 would be 2·776978, $m_2 = 0{\cdot}409833$, $a_1 = 13{\cdot}52728$, and $a_2 = 1{\cdot}99638$, and the graduation would be the same, but the numbers in the columns of the table on page 55 would run in the opposite order.

8. The arithmetical work is usually heavy. There is no royal road to success in it except care, system and the use of common-sense at the final stage. It is irritating at the end of a lengthy piece of arithmetic to find a slip at an early stage and to have to recalculate, but these slips become fewer and of less importance with experience, for when we are in practice we suspect a large error immediately an erroneous value is reached. Personally I use seven-figure logarithms as a rule and put a check on every step, although not necessarily to the last figure. This plan was followed with the arithmetical work in this book. The check might not disclose a slip which did not affect the graduation or only affected the final figures of a constant or coefficient. Thus if the last three figures of log y_0 on page 54 were wrong (which I have no reason to suppose) the mistake would be regrettable, but the graduation in the table on the following page would be practically unaffected. Moreover, difficulty may be found in reproducing exactly the numerical result of another calculator, owing to the usual unreliability of the end figures when many operations have been made. In lengthy arithmetic the two final

* In the formulae, and also in the diagrams at the end of the book, μ_3 (and the skewness) has been taken positive (or zero).

figures may be unreliable and two arithmetical processes may both be correct and yet give divergencies. This does not mean that five-figure logarithms are as good as seven, for if seven figures give five figures accurately, we assume that generally speaking five figure work will only be reliable to three figures.

FORMULAE FOR MOMENTS

THESE FORMULAE APPLY TO ALL THE TYPES OF CURVES

$$\left.\begin{aligned} \nu_1' &= d \\ \nu_2 &= \nu_2'-d^2 \\ \nu_3 &= \nu_3'-3d\nu_2-d^3 \\ \nu_4 &= \nu_4'-4d\nu_3-6d^2\nu_2-d^4 \end{aligned}\right\} \text{ or } \left\{\begin{aligned} \nu_1' &= d \\ \nu_2 &= v_2'-d^2 \\ \nu_3 &= \nu_3'-3d\nu_2'+2d^3 \\ \nu_4 &= \nu_4'-4d\nu_3'+6d^2\nu_2'-3d^4 \end{aligned}\right.$$

or $S_2 = d$

$$\nu_2 = 2S_3-d(1+d)$$
$$\nu_3 = 6S_4-3\nu_2(1+d)-d(1+d)(2+d)$$
$$\nu_4 = 24S_5-2\nu_3\{2(1+d)+1\}-\nu_2\{6(1+d)(2+d)-1\} \\ -d(1+d)(2+d)(3+d)$$

$$\left.\begin{aligned} \mu_2 &= \nu_2-\tfrac{1}{12} \\ \mu_3 &= \nu_3 \\ \mu_4 &= \nu_4-\tfrac{1}{2}\nu_2+\tfrac{7}{240} \end{aligned}\right\}$$ Sheppard's adjustments when the curve has high contact at both ends

$$\sigma \text{ (standard deviation)} = \sqrt{\mu_2}$$
$$\beta_1 = \mu_3^2/\mu_2^3$$
$$\beta_2 = \mu_4/\mu_2^2$$
$$\kappa = \frac{\beta_1(\beta_2+3)^2}{4(4\beta_2-3\beta_1)(2\beta_2-3\beta_1-6)}$$

FIRST MAIN TYPE (TYPE I)

$$y = y_0\left(1+\frac{x}{a_1}\right)^{m_1}\left(1-\frac{x}{a_2}\right)^{m_2} \qquad (-a_1 < x < a_2)$$

where $m_1/a_1 = m_2/a_2$

Origin at mode (antimode)

The values to be calculated in order are

$$r = 6(\beta_2-\beta_1-1)/(6+3\beta_1-2\beta_2)$$
$$a_1+a_2 = \tfrac{1}{2}\sqrt{\mu_2}\sqrt{\{\beta_1(r+2)^2+16(r+1)\}}$$

The m's are given by

$$\frac{1}{2}\left\{r-2\pm r(r+2)\sqrt{\frac{\beta_1}{\beta_1(r+2)^2+16(r+1)}}\right\}$$

(when μ_3 is positive, m_2 is the positive root)

$$y_0 = \frac{N}{a_1+a_2}\cdot\frac{m_1^{m_1}m_2^{m_2}}{(m_1+m_2)^{m_1+m_2}}\cdot\frac{\Gamma(m_1+m_2+2)}{\Gamma(m_1+1)\Gamma(m_2+1)}$$

$$\text{Mode} = \text{Mean}-\frac{1}{2}\cdot\frac{\mu_3}{\mu_2}\cdot\frac{r+2}{r-2}$$

If expressing curve with origin at mean (see Table 6 on page 45).

$$A_1+A_2 = a_1+a_2$$

$$(m_1+1)/A_1 = (m_2+1)/A_2$$

$$y_e = \frac{N}{A_1+A_2}\cdot\frac{(m_1+1)^{m_1}(m_2+1)^{m_2}}{(m_1+m_2+2)^{m_1+m_2}}\cdot\frac{\Gamma(m_1+m_2+2)}{\Gamma(m_1+1)\Gamma(m_2+1)}$$

For table of Γ functions see page 208, or tables described in Appendix 2. Proportion less than $X(-a_1 < X < a_2)$ is the incomplete beta function

$$I_{(X+a_1)/(a_2+a_1)}(m_1+1,\ m_2+1)$$

Notes

The usual shape of the curve is like that of the following example, but if m_1 and m_2 are approximately equal it is nearly symmetrical, if m_1 and m_2 are not small it tails off at both ends, and if both m_1 and m_2 are small it rises abruptly at both ends. If m_1 is negative and m_2 positive the curve is **J**-shaped; it starts at an infinite ordinate, falls rapidly and runs out at $x = a_2$ (for numerical example see p. 108). And conversely, if $m_1 > 0$ and $m_2 < 0$ we have a reversed **J**-shaped curve. If both m_1 and m_2 are negative, the curve is **U**-shaped, starting and ending with infinite ordinates and having an anti-mode instead of a mode as the usual origin.

In the **J**- and **U**-shaped curves, though the limiting ordinate is infinite, the area is finite. Care is needed in these cases when taking out the Γ function for $\Gamma(t)$ is required when $t < 1$ and the tables give $\log\Gamma(1+t)$, i.e. $\log t+\log\Gamma(t)$. In the case of **J**-shaped curves it is best to use the form with origin at the mean or express the curve in the form $y'x^{m_1}(a_1+a_n-x)^{m_2}$ with the origin at the start of the curve and

$$y' = \frac{N}{(a_1+a_2)^{m_1+m_2+1}}\cdot\frac{\Gamma(m_1+m_2+2)}{\Gamma(m_1+1)\Gamma(m_2+1)}$$

An interesting variant of the **J**-shaped curve arises when m_1 and m_2 are both arithmetically less than unity and one of them is negative. The shape is then like that of No. (11) in the diagram of curves at the end of the book, i.e. it is of twisted **J**-shape (for example and further notes, see pp. 94–95).

Example

As an example of this type the figures in Table 1 (Example 2) may be used. The moments were first found by the summation method (see Chapter 3, § 9) as shown in the following table. The reader can check the result by recalculating the moments by the more direct method, taking age 42 as the arbitrary origin. This is how I should myself usually do the work; I only use the summation method when the series is a very long one, and I give it here merely by way of example.

Central age of group	Exposed to risk Example 2 of Table 1	First sum	Second sum	Third sum	Fourth sum
17	34	1000	5 175	19 809	64 389
22	145	966	4 175	14 634	44 580
27	156	821	3 209	10 459	29 946
32	145	665	2 388	7 250	19 487
37	123	520	1 723	4 862	12 237
42	103	397	1 203	3 139	7 375
47	86	294	806	1 936	4 236
52	71	208	512	1 130	2 300
57	55	137	304	618	1 170
62	37	82	167	314	552
67	21	45	85	147	238
72	13	24	40	62	91
77	7	11	16	22	29
82	3	4	5	6	7
87	1	1	1	1	1
Totals	1000	5175	19 809	64 389	186 638

$$S_2 = 5175/1000 = 5{\cdot}175$$
$$S_3 = 19\,809/1000 = 19{\cdot}809$$
$$S_4 = 64\,389/1000 = 64{\cdot}389$$
$$S_5 = 186\,638/1000 = 186{\cdot}638$$

The next step is to find the moments about the centroid vertical using the formulae on page 51, and, in this case, as no adjustments * were made in the moments the ν's and μ's are the same.

$$\mu_2 = 7{\cdot}66237 \qquad \mu_3 = 15{\cdot}1069 \qquad \mu_4 = 172{\cdot}326$$
$$\beta_1 = 0{\cdot}5072955 \quad \beta_2 = 2{\cdot}935110$$

From the values of β_1 and β_2 the criterion (κ) can be calculated, and its value being $-0{\cdot}2645$ shows that Type I must be used (see Table 6).

$$\begin{array}{rlrl} r = & 5{\cdot}186811 & \log r = & 0{\cdot}7149004 \\ r+1 = & 6{\cdot}186811 & \log(r+1) = & 0{\cdot}7914669 \\ r+2 = & 7{\cdot}186811 & \log(r+2) = & 0{\cdot}8565363 \\ r-2 = & 3{\cdot}186811 & \log(r-2) = & 0{\cdot}5033563 \\ a_1+a_2 = & 15{\cdot}52366 & a_1 = & 1{\cdot}99638 \\ m_1 = & 0{\cdot}409833 & a_2 = & 13{\cdot}52728 \\ m_2 = & 2{\cdot}776978 & \text{Mean}-\text{mode} = & 2{\cdot}223116 \end{array}$$

It will be noted that the expression $\sqrt{\{\beta_1(r+2)^2+16(r+1)\}}$ occurs in both the values of (a_1+a_2) and m.

The mean is at age $12+5{\cdot}175\times 5 = 37{\cdot}8750$, and the mode at age $37{\cdot}8750-2{\cdot}223116\times 5 = 26{\cdot}75942$.

The skewness is 0·8032.

The calculation of $\log y_0$ is as follows:

$$\begin{aligned} \log N &= 3{\cdot}00000 \\ \operatorname{colog}(a_1+a_2) &= \bar{2}{\cdot}80901 \\ m_1 \log m_1 &= \bar{1}{\cdot}84123 \\ m_2 \log m_2 &= 1{\cdot}23179 \\ \operatorname{colog}(r-2)^{r-2} &= \bar{2}{\cdot}39590 \\ \log \Gamma(r) &= 1{\cdot}50406 \\ \operatorname{colog}\Gamma(m_1+1) &= 0{\cdot}05219 \\ \operatorname{colog}\Gamma(m_2+1) &= \bar{1}{\cdot}34037 \\ \log y_0 &= 2{\cdot}17455 \end{aligned}$$

where of course, $\log\Gamma(m_2+1) = \log\Gamma(3{\cdot}776978) = \log 2{\cdot}776978 + \log 1{\cdot}776978 + \log\Gamma(1{\cdot}776978)$, the last value being taken from the table at the end of the book.

* The moments should have been adjusted by one of the methods suitable when the curve is abrupt. These have been discussed since the example was prepared, and it was decided not to recalculate—see, however, Appendix I. Similar qualifications apply to a few of the other examples.

The work to this point gives as the curve for graduating the statistics

$$y = 149{\cdot}47\left\{1+\frac{x}{1{\cdot}99638}\right\}^{0{\cdot}409833}\left\{1-\frac{x}{13{\cdot}52728}\right\}^{2{\cdot}776978}$$

where the origin is at age 26·75942 (which is the modal value of x) and the unit is five years.

The following table shows the calculation of ordinates of the curve from the equation just given:

Age	$1+(x/a_1)$	$1-(x/a_2)$	log (2)	log (3)	m_1 × col. (4)	m_2 × col. (5)	col. (6) + col. (7) + log y_0 = log y_x	y_x	
(1)	(2)	(3)	(4)	(5)	(6)	(7)	(8)	(9)	(10)
17	0·02228	1·14429	$\bar{2}$·34792	0·05854	$\bar{1}$·3229	0·1626	1·6601	45·7	44
22	0·52319	1·07037	$\bar{1}$·71866	0·02955	$\bar{1}$·8847	0·0821	2·1404	138·2	137
27	1·02410	0·99644	0·01034	$\bar{1}$·99845	0·0042	$\bar{1}$·9957	2·1745	149·5	149
32	1·52501	0·92252	0·18327	$\bar{1}$·96498	0·0751	$\bar{1}$·9027	2·1525	142·1	142
37	2·02592	0·84859	0·30662	$\bar{1}$·92870	0·1257	$\bar{1}$·8020	2·1023	126·6	127
42	2·52683	0·77466	0·40257	$\bar{1}$·88911	0·1650	$\bar{1}$·6921	2·0317	107·6	108
47	3·02774	0·70074	0·48111	$\bar{1}$·84556	0·1972	$\bar{1}$·5711	1·9429	87·7	88
52	3·52865	0·62681	0·54760	$\bar{1}$·79714	0·2244	$\bar{1}$·4367	1·8357	68·5	69
57	4·02956	0·55289	0·60526	$\bar{1}$·74264	0·2481	$\bar{1}$·2853	1·7080	51·0	51
62	4·53047	0·47896	0·65615	$\bar{1}$·68030	0·2689	$\bar{1}$·1122	1·5557	36·0	36
67	5·03136	0·40504	0·70169	$\bar{1}$·60750	0·2876	$\bar{2}$·9100	1·3722	23·6	24
72	5·53229	0·33111	0·74291	$\bar{1}$·51997	0·3045	$\bar{2}$·6670	1·1461	14·0	14
77	6·03320	0·25719	0·78055	$\bar{1}$·41025	0·3199	$\bar{2}$·3623	0·8568	7·2	7
82	6·53411	0·18326	0·81519	$\bar{1}$·26307	0·3341	$\bar{3}$·9535	0·4622	2·9	3
87	7·03502	0·10934	0·84726	$\bar{1}$·03878	0·3472	$\bar{3}$·3307	$\bar{1}$·8525	0·7	1
92	7·53593	0·03541	0·87714	$\bar{2}$·54913	6·3595	$\bar{5}$·9709	$\bar{3}$·5050	...	...

Cols. (2) and (3) have a constant first difference, viz. $1/a_1$ or 0·500907, and $1/a_2$ or 0·073925. The value at any point having been calculated and checked, the other items are formed continuously. Cols. (4)–(9) explain themselves, but we may remark that it is generally advisable to use a larger number of figures than five in taking logarithms, especially if m_1 or m_2 is large. A little care is necessary in multiplying such numbers as $\bar{1}$·71866 by m_1(0·409833). Col. (10) gives the area, and was formed by applying one of the formulae on page 47. The area of the first group must be treated separately, as the curve starts at age 16·7775, and the base of the group is therefore 2·7225 in length, instead of 5 years as in the other cases. A good way to find the area is to calculate the ordinates for the middle and ends of the base, and apply Simpson's rule, remembering to multiply the result by 2·7225/5 to allow for the different length of the base.

The mid-ordinate is 92·1, the ordinate at the end of the base is 116·5, and the ordinate at the start is of course zero; the area is approximately *

$$\frac{2\cdot7225}{5}\times\tfrac{1}{6}\{0+4\times92\cdot1+116\cdot5\} = 44$$

Some people find it better when calculating the ordinates to use the form given in the Notes on page 52, with the origin at the start of the curve; it avoids bringing in the reciprocals of a_1 and a_2. The columns of $\log x$ and $\log(a_1+a_2-x)$ can be formed continuously. The initial values will have to be calculated and as a check one or perhaps two other values.

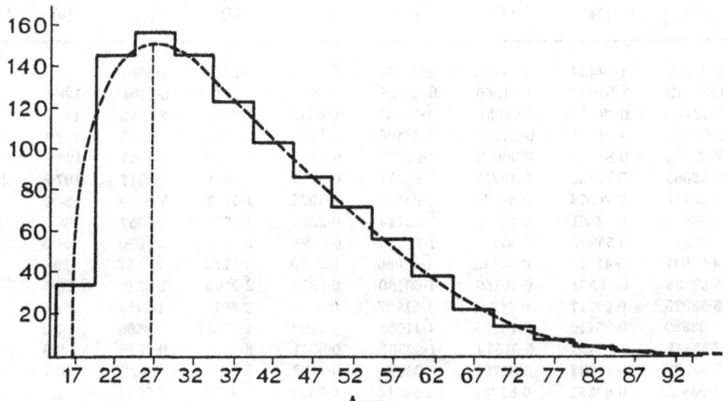

TYPE I

Proof of Formulae †

The equation to the curve is $y=y_0\left(1+\frac{x}{a_1}\right)^{m_1}\left(1-\frac{x}{a_2}\right)^{m_2}$, where $m_1/a_1 = m_2/a_2$.

Let $a_1+a_2 = b$ and $z = (a_1+x)/(a_1+a_2)$.

The area from $x = -a_1$ to $x = +a_2$ is the frequency N.

$$\therefore \quad N = \int_{-a_1}^{a_2} y_0\left(1+\frac{x}{a_1}\right)^{m_1}\left(1-\frac{x}{a_2}\right)^{m_2} dx$$

$$= \int_{-a_1}^{a_2} \frac{y_0}{a_1^{m_1}a_2^{m_2}}(a_1+x)^{m_1}(a_2-x)^{m_2}\,dx$$

* For greater accuracy use more ordinates or Tables of incomplete B-functions.

† The reader who has little acquaintance with formulae of reduction and the Γ and B-functions, should consult Appendix 2 before reading the proofs of the formulae for this and the other types.

$$= \int_0^1 \frac{y_0}{a_1^{m_1}a_2^{m_2}}[z(a_1+a_2)]^{m_1}[(1-z)(a_1+a_2)]^{m_2}(a_1+a_2)\,dz$$

$$= \int_0^1 \frac{y_0(a_1+a_2)^{m_1+m_2+1}}{a_1^{m_1}a_2^{m_2}} z^{m_1}(1-z)^{m_2}\,dz$$

$$= \frac{y_0(m_1+m_2)^{m_1+m_2}(a_1+a_2)}{m_1^{m_1}m_2^{m_2}} B(m_1+1,\, m_2+1)$$

Or $$y_0 = \frac{N}{b}\cdot\frac{m_1^{m_1}m_2^{m_2}}{(m_1+m_2)^{m_1+m_2}}\cdot\frac{\Gamma(m_1+m_2+2)}{\Gamma(m_1+1)\Gamma(m_2+1)}$$

Using the same method for the moments as that just given for the area, we see that the nth moment, about the line parallel to the axis of y through $x = -a_1$, is

$$N\mu_n' = \int_{-a_1}^{a_2} \frac{y_0}{a_1^{m_1}a_2^{m_2}}(a_1+x)^n(a_1+x)^{m_1}(a_2-x)^{m_2}\,dx$$

$$= \int_0^1 \frac{y_0(a_1+a_2)^{m_1+m_2+n+1}}{a_1^{m_1}a_2^{m_2}} z^{m_1+n}(1-z)^{m_2}\,dz$$

$$= \frac{y_0(m_1+m_2)^{m_1+m_2}b^{n+1}}{m_1^{m_1}m_2^{m_2}}\cdot\frac{\Gamma(m_1+n+1)\Gamma(m_2+1)}{\Gamma(m_1+m_2+n+2)}$$

Now, since $\Gamma(p) = (p-1)\Gamma(p-1)$, the moments about the line parallel to the axis of y through $x = -a_1$ are as follows:

$$\mu_1' = \frac{b(m_1+1)}{m_1+m_2+2}$$

$$\mu_2' = \frac{b^2(m_1+1)(m_1+2)}{(m_1+m_2+2)(m_1+m_2+3)} \quad \text{and so on.}$$

Changing the origin in order to get moments about the mean and writing $m_1' = m_1+1$ and $m_2' = m_2+1$ and $r = m_1'+m_2'$, we have

$$\mu_2 = \frac{b^2m_1'm_2'}{r^2(r+1)}$$

$$\mu_3 = \frac{2b^3m_1'm_2'(m_2'-m_1')}{r^3(r+1)(r+2)}$$

$$\mu_4 = \frac{3b^4m_1'm_2'\{m_1'm_2'(r-6)+2r^2\}}{r^4(r+1)(r+2)(r+3)}$$

We can simplify these expressions to obtain the equations on page 51 by writing $\beta_1 = \mu_3^2/\mu_2^3$, $\beta_2 = \mu_4/\mu_2^2$, and $\epsilon = m_1'm_2'$; then

$$\beta_1 = \frac{4(r^2-4\epsilon)(r+1)}{\epsilon(r+2)^2} \quad \text{or} \quad \frac{\beta_1(r+2)^2}{4(r+1)} = \frac{r^2}{\epsilon}-4$$

and $$\beta_2 = \frac{3(r+1)\{2r^2+\epsilon(r-6)\}}{\epsilon(r+2)(r+3)}$$

$$\text{or} \quad \frac{\beta_2(r+2)(r+3)}{3(r+1)} = \frac{2r^2}{\epsilon}+r-6$$

Eliminating r^2/ϵ we find

$$\frac{\beta_1(r+2)^2}{2(r+1)} - \frac{\beta_2(r+2)(r+3)}{3(r+1)} = -8-r+6$$

Dividing out by $r+2$ we have

$$r = \frac{6(\beta_2-\beta_1-1)}{3\beta_1-2\beta_2+6}$$

From the equation $\dfrac{\beta_1(r+2)^2}{4(r+1)} = \dfrac{r^2}{\epsilon}-4$ we have

$$\epsilon = \frac{r^2}{4+\frac{1}{4}\beta_1\dfrac{(r+2)^2}{r+1}}$$

and from the equation for μ_2

$$b^2 = \frac{\mu_2(r+1)r^2}{\epsilon}$$

The other equations follow at once from $r = m_1'+m_2'$ and $\epsilon = m_1'm_2'$. The distance between the mode and mean is

$$a_1-\mu_1' = a_1-bm_1'/(m_1'+m_2'),$$

which can be easily reduced to the form given. A general value (regardless of type) for the distance was given in Chapter 4, § 5.

SECOND MAIN TYPE (TYPE IV)

$$y = y_0\left(1+\frac{x^2}{a^2}\right)^{-m} e^{-\nu \tan^{-1} x/a}$$

Origin is $\nu a/r$ above mean

The values to be calculated in order are

$$r = \frac{6(\beta_2-\beta_1-1)}{2\beta_2-3\beta_1-6}$$

$$m = \tfrac{1}{2}(r+2)$$

$$\nu = \frac{-r(r-2)\sqrt{\beta_1}}{\sqrt{\{16(r-1)-\beta_1(r-2)^2\}}}$$

$$a = \sqrt{\frac{\mu_2}{16}}\sqrt{\{16(r-1)-\beta_1(r-2)^2\}}$$

$$y_0 = \frac{N}{aF(r, \nu)}$$

$$\text{Mode} = \text{mean} - \frac{1}{2}\frac{\mu_3(r-2)}{\mu_2(r+2)}$$

(Note that, since $2\beta_2-3\beta_1-6>0$ for Type IV, r must be greater than 3.)

Notes

The curve is skew and has unlimited range in both directions.

μ_3 and ν have opposite signs, i.e. when μ_3 is positive ν is negative.

A simple way to calculate the curve is to put it in the form

$$x = a\tan\theta, \qquad y = y_0 \cos^{r+2}\theta\, e^{-\nu\theta}$$

Then θ is taken as 10°, 20°, 30°, etc., and x and y found; this gives corresponding values of x and y, but the values of x will not be at equal intervals. (Note that in calculating $e^{-\nu\theta}$ the value of θ must be taken in circular measure.) If equidistant ordinates are required to be calculated accurately, it is probably as easy to use the first form, especially if tables of $\log(1+x^2)$ and $\tan^{-1}x$, are available.* The calculation and meaning of $F(r, \nu)$ are dealt with in the proof. The log of this function is tabulated in *Tables for Statisticians.* When r is fairly large a close approximation to y_0, where $\tan\phi = \nu/r$, is given by

$$\frac{N}{a}\cdot\sqrt{\frac{r}{2\pi}}\cdot\frac{e^{\frac{\cos^2\phi}{3r}-\frac{1}{12r}-\phi\nu}}{(\cos\phi)^{r+1}}$$

We appear to reach the expression that looks shortest and simplest with the origin as shown on the previous page; it has generally been used and it is therefore given. This origin has, however, no physical meaning and there is much to be said for using the more complicated looking form with the origin at the mean, namely

$$y = y_0\left\{1+\left(\frac{x}{a}-\frac{\nu}{r}\right)^2\right\}^{-m} e^{-\nu\tan^{-1}(x/a-\nu/r)}$$

(see table on page 45).

The value of this expression when $x = 0$, i.e. the value of the ordinate at the mean, is

$$y_0\left\{1+\frac{\nu^2}{r^2}\right\}^{-m} e^{-\nu\tan^{-1}(-\nu/r)} = \frac{N}{\sigma}\cdot\frac{1}{H(r, \nu)}$$

* Such tables are included in *Tracts for Computers*, XXIII, by L. J. Comrie.

where $H(r, \nu)$ is a function related to $F(r, \nu)$. Its logarithm is also tabulated in *Tables for Statisticians*. The reader will appreciate at once that this curve needs considerable care; it is the most difficult of all the Pearson-type curves.

Examples

The numbers in the following nearly symmetrical distribution represent the exposed to risk of sickness by Sutton's Sickness Tables (males—all durations) when the number of weeks' sickness is represented by the normal curve of error.

Central age	No. exposed	Type IV	Normal	S_U
5	10	6 *	3 *	6 *
10	13	15	11	15
15	41	46	43	45
20	115	129	137	130
25	326	313	349	313
30	675	646	718	646
35	1113	1110	1187	1109
40	1528	1550	1581	1550
45	1692	1733	1695	1734
50	1530	1535	1462	1536
55	1122	1074	1016	1074
60	610	596	638	595
65	255	265	186	265
70	86	96	93	96
75	26	29	27	29
80	8	8	6	8
85	2	2	1	2
90	1	1	0	0
95	1	0	0	0
...	9154	9154	9153	9153

* This group has been taken as the area of the rest of the curve.

The following values were obtained:

Mean = 44·577234

$\mu_2 = 4{\cdot}527608 \quad \mu_3 = -0{\cdot}705687 \quad \mu_4 = 65{\cdot}042264$

$\beta_1 = 0{\cdot}0053656 \quad \beta_2 = 3{\cdot}172912 \quad k = 0{\cdot}0128$

5-year units.

Type IV was used because, as there is a large number of cases, the standard error of κ will be small (see Chapter 9)

$$r = 39{\cdot}44254$$
$$\nu = 4{\cdot}388794 \text{ (positive because } \mu_3 \text{ is negative)}$$
$$a = 13{\cdot}11198$$
$$m = 20{\cdot}72127$$
$$\text{Sk.} = -0{\cdot}03309$$

When the 5-years unit with which we have been working is changed to one year, a becomes 65·5599, and $a^2 = 4298{\cdot}100$

$$\text{The origin} = \text{mean} + \nu a/r$$
$$= 51{\cdot}872124$$

The mode, which is wanted if the curve is drawn, is at 44·92928.

As r is large the approximate form for y_0 was used,

$$\tan\phi = 4{\cdot}388794/39{\cdot}44254$$

whence $\phi = 0{\cdot}110785$ radians; hence $\log\cos\phi = \bar{1}{\cdot}9973293$, and from this y_0 is found to be 277·6026.

The value was checked by the tables in *Tables for Statisticians*.

The calculation of ordinates by the double process is as follows:

θ	$x =$ (age $-51{\cdot}87212$)	$-4{\cdot}388794\ \theta \log e$	$41{\cdot}42254 \log\cos\theta$	$\log y$	y
0°	0	...	...	2·44342	277·60
1°	1·1443	$\bar{1}{\cdot}96673$	$\bar{1}{\cdot}99726$	2·40741	255·51
2°	2·2887	$\bar{1}{\cdot}93347$	$\bar{1}{\cdot}98903$	2·36593	232·23

The second column is formed directly from the tables of $\tan\theta$ by multiplying by a, and as x is required in years, $13{\cdot}11198 \times 5 = 65{\cdot}5599$ should be used for a. The fourth column is formed by multiplying $\log\cos\theta$ by $r+2$, and the third continuously by addition. When θ is negative, the third column has to be subtracted from the fourth; i.e. it ceases to be negative and becomes positive. In each case the fifth is formed from the fourth+the third+$\log y_0$.

When drawing a curve of this type the position and height of the mode ($x = -6{\cdot}9428$, $y = 350{\cdot}07$) can be noted and then corresponding points inserted, e.g. $x = +1{\cdot}1443$ and $y = 255{\cdot}51$. Care must be taken to give the curve its maximum at the right point.

If the calculation is made directly, the following columns can be used:

x/a	$1+x^2/a^2$	$\log(1+x^2/a^2)$	$\tan^{-1} x/a$ in degrees, etc.	col. (4) in circular measure	col. (5) $\times(-\nu \log_{10} e)$	$-m \times$ col. (3)	$\log y_0$ +(6)+(7)	$y =$ antilog (8)
(1)	(2)	(3)	(4)	(5)	(6)	(7)	(8)	(9)
...	...	...	...	...	...	...	...	...
...	...	...	...	...	...	...	...	...

Col. (2) can be formed by differences since $\Delta(1+X^2) = 2X+1$; $\tan^{-1} x/a$ has to be found by using a table of the tangents of angles inversely. A table helpful for obtaining col. (5) from col. (4) will be found in *Chambers' Mathematical Tables* or in *Tables for Statisticians*, for example.

The troublesome work of inverse interpolation in degrees, minutes and seconds can be avoided by numbering the items in a table of $\tan \theta$ from 0 onwards. *Chambers' Tables*, for instance, give tangents for each minute in the following form:

′	0°	1°	2°, etc.
0	0·0000000	(60)0·0174551	(120)0·0349208
1	0·0002909	(61)0·0177460	(121)0·0352120
2	0·0005818	(62)0·0180370	(122)0·0355033
3	0·0008727	(63)0·0183280	(123)0·0357945

If in the column headed 1° we insert 60, 61, etc., and in the column headed 2° we insert 120, 121, etc., as indicated by the figures in brackets, we can make the inverse interpolation in minutes. Then, as one minute is 0·0002908882 radians, we can obtain the figure we require by multiplying by the conversion factor. In practice however it would be combined into one multiplier with $(-\nu \log_{10} e)$ and col. (6) would be found directly from col. (4) by multiplying in our example, by 0·003519003. The labour of inserting the minutes in a printed table is small, as all we need to do is to write the number of minutes under the number of degrees at the head of each column and add thereto at sight the marginal minutes when the interpolations are being made. Of course, if tables with radian arguments are used, these difficulties do not arise.

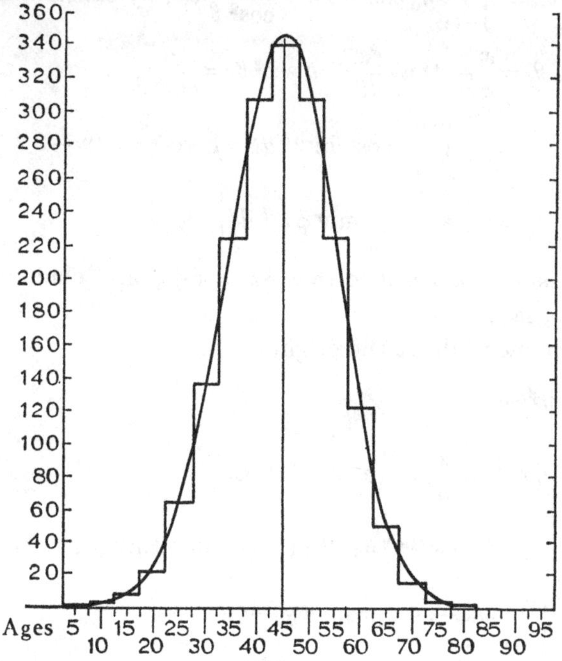

TYPE IV

The last two columns in the table on page 60 give results of graduation of the data by a normal and a S_U curve respectively. The latter will be discussed in Chapter 7. In regard to the former, it is notable that although the original distribution is nearly symmetrical, the normal curve gives a markedly worse fit than either of the other two curves.

Proof

In $$y = y_0\left\{1+\frac{x^2}{a^2}\right\}^{-m} e^{-\nu \tan^{-1} x/a} \text{ put } \tan\theta = \frac{x}{a}$$

$\therefore$ $$\theta = \tan^{-1}\frac{x}{a}$$

and $$\left\{1+\left(\frac{x}{a}\right)^2\right\}^{-m} = \{1+\tan^2\theta\}^{-m} = (\sec^2\theta)^{-m} = \cos^{2m}\theta$$

$$\therefore \qquad y = y_0 \cos^{2m}\theta\, e^{-\nu\theta}$$

Now
$$N = \int_{-\infty}^{+\infty} y_0\left\{1+\frac{x^2}{a^2}\right\}^{-m} e^{-\nu\tan^{-1}x/a}\,dx$$

$$= \int_{-\frac{1}{2}\pi}^{\frac{1}{2}\pi} y_0 \cos^{2m}\theta\, e^{-\nu\theta}\frac{a}{\cos^2\theta}\,d\theta, \text{ by substituting}$$

$$\tan\theta = \frac{x}{a} \text{ so that } \frac{dx}{d\theta} = a\sec^2\theta = \frac{a}{\cos^2\theta}$$

$$= y_0 a\int_{-\frac{1}{2}\pi}^{\frac{1}{2}\pi} \cos^r\theta\, e^{-\nu\theta}\,d\theta \text{ where } r = 2m-2$$

$$= y_0 a e^{-\frac{1}{2}\nu\pi}\int_0^{\pi} \sin^r\phi\, e^{\nu\phi}\,d\phi,$$

substituting $\sin\phi$ for $\cos\theta$ so that $\frac{1}{2}\pi = \theta+\phi$ and changing limits, $= y_0 aF(r,\nu)$, say.

The nth moment about the origin is

$$\mu_n' = \frac{1}{N}\int_{-\infty}^{\infty} yx^n\,dx$$

$$= \frac{1}{N}\int_{-\infty}^{\infty} y_0 x^n\left\{1+\frac{x^2}{a^2}\right\}^{-m} e^{-\nu\tan^{-1}x/a}\,dx$$

$$= \frac{1}{N}\int_{-\frac{1}{2}\pi}^{\frac{1}{2}\pi} y_0 a^{n+1}\cos^{2m-2}\theta\, e^{-\nu\theta}\,d\theta, \text{ by substituting as above,}$$

$$= \frac{y_0 a^{n+1}}{N}\int_{-\frac{1}{2}\pi}^{\frac{1}{2}\pi} \cos^{r-n}\theta \sin^n tan^n\theta\, e^{-\nu\theta}\,d\theta$$

$$= \frac{y_0 a^{n+1}}{N}\left[-\frac{\cos^{r-n+1}\theta\, e^{-\nu\theta}}{r-n+1}\right.$$
$$\left. + \int\left\{\frac{\cos^{r-n+1}\theta}{r-n+1}[\sin^{n-2}\theta\cos\theta\, e^{-\nu\theta}(n-1) - \nu\, e^{-\nu\theta}\sin^{n-1}\theta]\right\}d\theta\right]_{-\frac{1}{2}\pi}^{\frac{1}{2}\pi}$$

by integrating by parts and treating $\sin^{n-1}\theta\, e^{-\nu\theta}$ as one part and $\cos^{r-n}\theta\sin\theta$ as the other, and remembering that

$$\int \cos^{r-n}\theta\sin\theta\,d\theta = -\frac{\cos^{r-n+1}\theta}{r-n+1}.$$

Now, since $\cos^{r-n+1}\theta\sin^{n-1}\theta\, e^{-\nu\theta} = 0$ when θ becomes $\frac{1}{2}\pi$ or $-\frac{1}{2}\pi$, we have

$$\mu_n' = \frac{y_0 a^{n+1}}{N(r-n+1)}\int_{-\frac{1}{2}\pi}^{\frac{1}{2}\pi}\{(n-1)\cos^{r-n+2}\theta\sin^{n-2}\theta\, e^{-\nu\theta}$$
$$-\nu\cos^{r-n+1}\theta\sin^{n-1}\theta\, e^{-\nu\theta}\}\,d\theta$$

$$= \frac{a}{r-n+1}\{(n-1)a\mu_{n-2}' - \nu\mu_{n-1}'\} \qquad (n<2m-1)$$

Further,

$$\mu_1' = \frac{y_0 a^2}{N}\int_{-\frac{1}{2}\pi}^{\frac{1}{2}\pi} \cos^r\theta \tan\theta\, e^{-\nu\theta}\,d\theta$$

$$= \frac{y_0 a^2}{Nr}\left\{-\int_{-\frac{1}{2}\pi}^{\frac{1}{2}\pi} \nu \cos^r\theta\, e^{-\nu\theta}\,d\theta\right\}$$

by putting $n = 1$ in the above equation for μ_n'

$$= -\frac{\nu a}{r}, \text{ because } N = y_0 a\int_{-\frac{1}{2}\pi}^{\frac{1}{2}\pi} \cos^r\theta\, e^{-\nu\theta}\,d\theta \qquad (m>1)$$

Using the last result with the formula for the nth in terms of the two previous moments, and remembering that μ_0' is unity,

$$\mu_2' = \frac{a^2}{r(r-1)}(r+\nu^2) \qquad (m>\tfrac{3}{2})$$

$$\mu_3' = -\frac{a^3\nu}{r(r-1)(r-2)}(3r-2+\nu^2) \qquad (m>2)$$

$$\mu_4' = \frac{a^4}{r(r-1)(r-2)(r-3)}\{3r(r-2)+\nu^2(6r-8)+\nu^4\} \qquad (m>\tfrac{5}{2})$$

Referring these moments to the centroid vertical, we have, by putting $d = \mu_1' = -(a\nu/r)$ in the formulae on page 51,

$$\mu_2 = \frac{a^2}{r^2(r-1)}(r^2+\nu^2) \qquad (m>\tfrac{3}{2})$$

$$\mu_3 = -\frac{4a^3\nu(r^2+\nu^2)}{r^3(r-1)(r-2)} \qquad (m>2)$$

$$\mu_4 = \frac{3a^4(r^2+\nu^2)\{(r+6)(r^2+\nu^2)-8r^2\}}{r^4(r-1)(r-2)(r-3)} \qquad (m>\tfrac{5}{2})$$

If now, we put z for $r^2+\nu^2$, and write as before,

$$\beta_1 = \frac{\mu_3^2}{\mu_2^3} \quad \text{and} \quad \beta_2 = \frac{\mu_4}{\mu_2^2}$$

we have

$$-\frac{\beta_1(r-2)^2}{2(r-1)} = \frac{8r^2}{z} - 8$$

and

$$\frac{\beta_2(r-2)(r-3)}{3(r-1)} = r+6-\frac{8r^2}{z}$$

Adding and dividing out by $r-2$, we have

$$r = \frac{6(\beta_2-\beta_1-1)}{2\beta_2-3\beta_1-6}$$

and

$$z = \frac{r^2}{1-\dfrac{\beta_1(r-2)^2}{16(r-1)}}$$

Finally, since $\nu^2 = z-r^2$, the other formulae on pages 58–59 follow at once.

Since the tangent at the top of the maximum ordinate is parallel to the axis of x, the position of the mode is such that dy/dx is zero at that point, i.e.

$$y_0\left\{1+\frac{x^2}{a^2}\right\}^{-(m+1)} e^{-\nu \tan^{-1} x/a}\left[-\frac{2mx}{a^2}-\frac{\nu}{a}\right]$$

is zero. There are three cases, $x = -\infty$, $x = +\infty$, and a value of x such that $(2mx/a^2)+(\nu/a)$ is zero, or $x = -(\nu a/2m)$. The distance of the mean from the origin is μ_1 or $-(\nu a/r)$, and therefore, the distance between the mean and mode is $-[2\nu a/r(r+2)]$, which reduces to the expression given on page 59, when the values for ν and a are inserted.

It will be useful to give another example of the calculation of y_0 for curves of this type, and we may take a curve where $r = 29{\cdot}590$, $\nu = 19{\cdot}886$, $a = 13{\cdot}650$, $N = 2162$. Hence $\tan\phi = 0{\cdot}67205$, $\phi = 33°\ 54'\tfrac{8}{42}$, $\cos\phi = 0{\cdot}82998$, $\log\cos\phi = \bar{1}{\cdot}91907$, and ϕ in radians is $0{\cdot}59172$.

$$\begin{aligned}
\log N &= 3{\cdot}33486\\
\operatorname{colog} a &= \bar{2}{\cdot}86486\\
\tfrac{1}{2}\log r &= 0{\cdot}73557\\
\log\frac{1}{\sqrt{2\pi}} &= \bar{1}{\cdot}60091
\end{aligned}$$

$$\begin{aligned}
\frac{\cos^2\phi}{3r} &= \quad 0{\cdot}00776\\
-\frac{1}{12r} &= -\ 0{\cdot}00282\\
-\phi\nu &= \underline{-11{\cdot}76700}\\
& \quad -11{\cdot}762 \times \log_{10} e = \bar{6}{\cdot}89183\\
& \operatorname{colog}(\cos\phi)^{r+1} = \underline{2{\cdot}47564}\\
& \qquad\qquad \underline{\bar{1}{\cdot}90367}\\
& \qquad\qquad y_0 = 0{\cdot}80107
\end{aligned}$$

The form just considered is sufficiently accurate for all practical purposes provided ν is not very small. If $\nu < 2$ the tables in *Tables for Statisticians* must be used.

THIRD MAIN TYPE (TYPE VI)

$y = y_0(x-a)^{q_2}x^{-q_1}$

Origin at a before start of curve

The values to be calculated in order are

$$r = \frac{6(\beta_2-\beta_1-1)}{6+3\beta_1-2\beta_2}$$

$$a = \tfrac{1}{2}\sqrt{\mu_2}\sqrt{\{\beta_1(r+2)^2+16(r+1)\}}$$

q_2 and $-q_1$ are given by

$$\frac{r-2}{2} \pm \frac{r(r+2)}{2}\sqrt{\frac{\beta_1}{\beta_1(r+2)^2+16(r+1)}}$$

$$y_0 = \frac{Na^{q_1-q_2-1}\,\Gamma(q_1)}{\Gamma(q_1-q_2-1)\Gamma(q_2+1)}$$

$$\text{Origin} = \text{Mean} - \frac{a(q_1-1)}{q_1-q_2-2}$$

$$\text{Mode} = \text{Mean} - \frac{1}{2}\cdot\frac{\mu_3}{\mu_2}\cdot\frac{r+2}{r-2}$$

If expressing curve with origin at mean (see Table 6, on page 45):

$$A_1 = \frac{a(q_1-1)}{(q_1-1)-(q_2+1)},\quad A_2 = \frac{a(q_2+1)}{(q_1-1)-(q_2+1)}$$

$$y_e = \frac{N(q_2+1)^{q_2}(q_1-q_2-2)^{q_1-q_2}\Gamma(q_1)}{a(q_1-1)^{q_1}\Gamma(q_1-q_2-1)\Gamma(q_2+1)}$$

Notes

The range is from a to ∞ and the method is like that of Type I. If μ_3 is negative, then a is negative and the range is from $-\infty$ to a.

r is always negative and q_1 is greater than q_2. If q_2 is negative, the curve is J-shaped.

The reader will probably find it easier to work with the origin at the mean, and in the numerical example both forms are shown.

Example

The number of entrants, limited payment policies, 1863–93 experience was summed in groups of ten years of age and divided by 100, and the following series was obtained:

No. of entrants ÷100	Graduated by Type VI curve
1	1
56	50
167	168
98	100
34	36
9	10
2	2
1	0·5
368	368

The moments, etc. were

Mean at 0·402174 after the centre of 167 group

$\mu_2 =$ 0·928835 $\quad \mu_3 =$ 0·893096 $\quad \mu_4 =$ 4·088800

$\beta_1 =$ 0·9953605 $\quad \beta_2 =$ 4·739349 $\quad \kappa =$ 1·895

$r = -33{\cdot}42129 \quad 1-q_1 = -41{\cdot}03080 \quad 1+q_2 = 7{\cdot}60950$

$q_1 =$ 42·03080 $\quad q_2 =$ 6·60950 $\quad a =$ 10·37947

$\log y_0 =$ 46·1821

The origin is 12·74270 before the mean or 12·34053 before the centre of the 167 group, and the curve starts at 12·34053—10·37949 = 1·96106 before the centre of the largest group. This makes the start of the curve at about age 10, which is reasonable.

If we use the origin at the mean, we have

$$A_1 = 12{\cdot}74270, \qquad A_2 = 2{\cdot}36324, \qquad y_e = 147{\cdot}4$$

and the range is from $-2{\cdot}36324$ to ∞.

The curve was calculated as follows:

x	$\log x$	$\log(x-a)$	$-q_1 \log x$	$q_2 \log(x-a)$	$\log y$	y
(1)	(2)	(3)	(4)	(5)	(6)	(7)

There is no difficulty in writing down the values for cols (2) and (3) without using col. (1), as only the whole numbers in x and $x-a$

change, the decimal remaining constant so long as equidistant ordinates are required. Cols. (4) and (5) are obtained directly, and col. (6) by adding cols. (4) and (5) to log y_0.

The mode which is useful for drawing the curve is 0·02429 before the centre of the largest group.

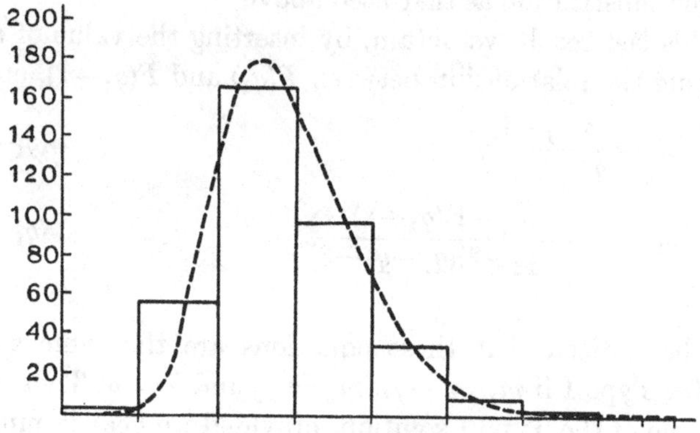

TYPE VI

With the origin at the mean the form of the columns is similar to that already shown for Type I.

Proof

$$N = \int_a^\infty y_0(x-a)^{q_2}x^{-q_1}dx$$

$$= \int_a^\infty y_0 a^{q_2-q_1}\left(\frac{x}{a}-1\right)^{q_2}\left(\frac{x}{a}\right)^{-q_1}dx$$

$$= \int_1^0 y_0 a^{q_2-q_1}\left(\frac{1}{z}-1\right)^{q_2} z^{q_1}(-az^{-2})dz$$

by substituting $1/z$ for x/a

$$= \int_0^1 y_0 a^{q_2-q_1+1}(1-z)^{q_2}z^{q_1-q_2-2}dz$$

$$y_0 = \frac{N}{a^{q_2-q_1+1}B(q_2+1,\, q_1-q_2-1)}$$

$$= \frac{N\Gamma(q_1)a^{q_1-q_2-1}}{\Gamma(q_2+1)\Gamma(q_1-q_2-1)}$$

The nth moment about the origin is

$$\mu'_n = \frac{1}{N}\int_a^\infty y_0 x^n (x-a)^{q_2} x^{-q_1} dx$$

$$= \frac{y_0}{N a^{q_1-q_2-n-1}} \frac{\Gamma(q_1-q_2-n-1)\Gamma(q_2+1)}{\Gamma(q_1-n)} \qquad (n<q_1-q_2-1)$$

by the same substitution as that used above.

From this last result we obtain, by inserting the value of y_0, and remembering the relationship between $\Gamma(q_1)$ and $\Gamma(q_1-1)$, etc.,

$$\mu'_1 = \frac{a(q_1-1)}{q_1-q_2-2} \qquad (q_1-q_2>2)$$

$$\mu'_2 = \frac{a^2(q_1-1)(q_1-2)}{(q_1-q_2-2)(q_1-q_2-3)} \qquad (q_1-q_2>3)$$

etc.

It will be noticed that these equations are the same as those obtained for Type I if $m_1 = -q_1$, $m_2 = q_2$ and $b = a$. Thus, we can use the whole of the Type I solution, provided we bear in mind that the range is from $x = a$ to $x = \infty$.

TRANSITION TYPES

'NORMAL CURVE OF ERROR'

$$y = y_0 e^{-x^2/c}$$

$$c = 2\sigma^2$$

$$y_0 = \frac{N}{\sqrt{(2\pi\mu_2)}}$$

$$Z(x) = (2\pi)^{-\frac{1}{2}} e^{-\frac{1}{2}x^2}$$

$$P(x) = \int_{-\infty}^{x} Z(x)dx$$

This curve has been known by various names, such as the Probability Curve and the Gaussian Curve. It was discussed before Gauss by de Moivre and Laplace. It is the limit of $(p+q)^n$ where $p+q = 1$, when n approaches infinity and if neither p nor q is very small. It gives a close representation of $(\frac{1}{2}+\frac{1}{2})^n$ even when n is not large.

Examples

The following table gives, in col. (2), the sums assured and bonuses, and in col. (4) the reserves resulting from grouping a number of Endowment Assurances according to their office years of birth:

Central age for groups of 5 years of birth	Sums assured and bonuses/1000		Reserves/1000	
	Ungraduated	Graduated	Ungraduated	Graduated
(1)	(2)	(3)	(4)	(5)
17	11	16	0·6	0·6
22	48	45	2·8	2·8
27	124	115	11·5	10·9
32	213	213	27·7	30·1
37	281	287	59·1	58·4
42	295	282	84·7	79·9
47	185	202	74·1	77·0
52	104	105	50·5	52·2
57	40	40	23·2	25·0
62	15	11	12·2	8·4
67	3	3	1·3	2·4
Total	1319	1319	347·7	347·7

The following table shows the moments and constants:

Constant	Sum assured and bonus	Reserves
Mean age	39·202426	43·967213
μ_2	3·066840	2·769635
μ_3	0·650127	0·029805
μ_4	27·02516	22·40663
β_1	0·014653	0·0000418
β_2	2·873346	2·920997
κ	−0·04	−0·0002
$\sigma(=\sqrt{\mu_2})$	1·751237	1·664222
σ^{-1}	0·5710248	0·6008813
y_0	300·4760	83·34959

The criteria for the normal curve are $\kappa = 0$, $\beta_1 = 0$, and $\beta_2 = 3$. The values given above do not differ very greatly from these, but a comparison of the graduated and ungraduated figures shows that the reserve curve agrees better than the sum assured curve; partly because the value of β_2 is closer to 3, and β_1 has a smaller value in the case of the reserves.

For the calculation of y_0 the value of

$$\operatorname{colog}\sqrt{2\pi} = \bar{1}\cdot6009100657$$

is required.

In finding the areas for the comparison between the graduated and ungraduated figures it is unnecessary to calculate the ordinates, as one of the calculated tables of the probability integral can be used. The table by W. F. Sheppard included in *Biometrika Tables* is very convenient, and the columns in the table below show how it was used to calculate the areas in one of the cases (the reserves). Sheppards' tables give the areas and ordinates of the normal curve in terms of

Age x	Distance from mean in calculation units, i.e. 5 years of age	y = Previous column $\times\,\sigma^{-1}$	Values of $P(y)$ from Sheppard's tables using differences (area up to y)	Difference of previous column = area for age group x to $x+5$	Area multiplied by 347·7 (total frequency)
14·5	−5·893443	−3·541258	...	0·00164 *	0·6
19·5	−4·893443	−2·940377	0·00164	0·00802	2·8
24·5	−3·893443	−2·339496	0·00966	0·03139	10·9
29·5	−2·893443	−1·738615	0·04015	0·08657	30·1
34·5	−1·893443	−1·137734	0·12762	0·16806	58·4
39·5	−0·893443	−0·536853	0·29568	0·22985	79·9
44·5	0·106557	0·064028	0·52553	0·22141	77·0
49·5	1·106557	0·664909	0·74694	0·15018	52·2
54·5	2·106557	1·265790	0·89712	0·07190	25·0
59·5	3·106557	1·866671	0·96902	0·02418	8·4
64·5	4·106557	2·467552	0·99320	0·00572	2·0
69·5	5·106557	3·068443	0·99892	0·00108 *	0·4

* Remainders of areas beyond 19·5 and 69·5.

the standard deviation; that is, he assumes the standard deviation to be unity, and his tables must be entered by using intervals of σ^{-1}. A short abstract from Sheppard's table is given on page 207. Some other published tables are based on the standard deviation multiplied by $\sqrt{2}$ and the distinction must be borne in mind if such tables are used.

The second column can be left out when the method has been grasped. The ages in the first column were taken consistently with the assumptions that 17, 22, etc., were the central ages of the groups.

If ordinates are required, the *Z* column in Sheppard's tables must be used. It was with its help that the curves in the figure were drawn. The statistics and curve for the reserves are shown by the dotted lines.

An average reserve for any group can be obtained by means of the graduated figures, and it could be used to test the reserves obtained at any future valuation. This is by no means the only

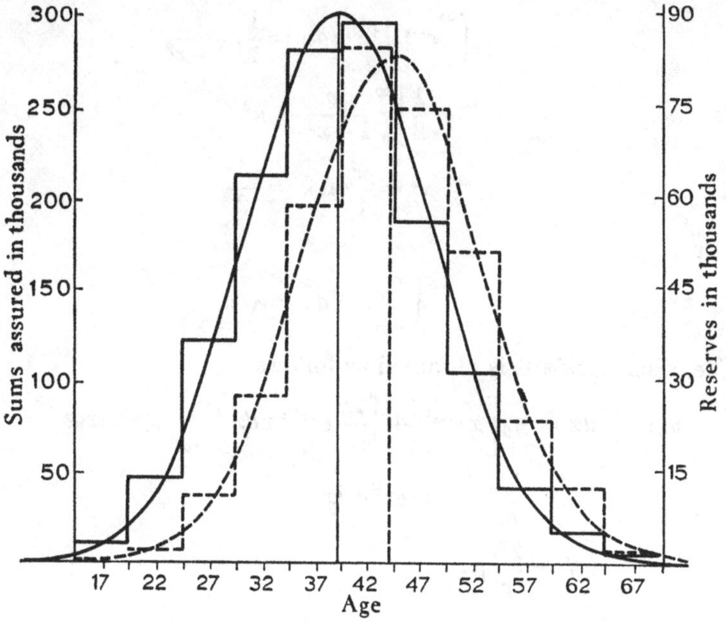

'NORMAL CURVE OF ERROR'

rough check that can be applied, but it is interesting because it shows a use to which frequency curves might be put in practical office routine.

Proof

To show that

$$\int_0^\infty e^{-x^2}\,dx = \frac{\sqrt{\pi}}{2}$$

let

$$\int_0^\infty e^{-x^2}\,dx = \kappa$$

then, substituting ax for x, we have

$$\int_0^\infty e^{-a^2x^2}a\,dx = \kappa$$

$$\therefore \qquad \int_0^\infty e^{-a^2(1+x^2)}a\,dx = \kappa\,e^{-a^2}$$

Hence

$$\int_0^\infty\int_0^\infty e^{-a^2(1+x^2)}a\,da\,dx = \kappa\int_0^\infty e^{-a^2}\,da = \kappa^2$$

But

$$\int_0^\infty e^{-a^2(1+x^2)}a\,da = \frac{1}{2}\cdot\frac{1}{1+x^2}$$

$\therefore$

$$\frac{1}{2}\int_0^\infty \frac{dx}{1+x^2} = \kappa^2$$

and

$$\kappa^2 = \frac{\pi}{4} \text{ or } \kappa = \frac{\sqrt{\pi}}{2}$$

Hence

$$\int_{-\infty}^\infty e^{-x^2}\,dx = \sqrt{\pi}$$

The other constant is obtained as follows:

$$\int_{-\infty}^\infty y_0\,e^{-x^2/c}\,dx = y_0\left[x\,e^{-x^2/c} + \int\frac{2x}{c}\,e^{-x^2/c}\,x\,dx\right]_{-\infty}^\infty \text{ by parts}$$

$$= \frac{2y_0}{c}\int_{-\infty}^\infty x^2\,e^{-x^2/c}\,dx$$

or

$$N = \frac{2N}{c}\mu_2$$

$\therefore$

$$c = 2\mu_2$$

TRANSITION TYPE (TYPE II)

$$y = y_0\left(1-\frac{x^2}{a^2}\right)^m$$

Origin at mode (= mean)

$$m = \frac{5\beta_2-9}{2(3-\beta_2)}$$

$$a^2 = \frac{2\mu_2\beta_2}{3-\beta_2}$$

$$y_0 = \frac{N\times\Gamma(2m+2)}{a\times 2^{2m+1}\{\Gamma(m+1)\}^2}$$

$$= \frac{N}{a\sqrt{\pi}}\cdot\frac{\Gamma(m+1\frac{1}{2})}{\Gamma(m+1)}$$

Notes and Proof

Put $\beta_1 = 0$ in Type I, for the curve is symmetrical, and therefore $\mu_3 = 0$. For the same reason it is clear that $m_1 = m_2$.

Approximations to Γ may be used if m is large.

If m is positive, the curve starts at zero, rises to a maximum and falls again to zero; but if m is negative, it starts at infinity, falls, and then rises to infinity again.

Example

In the discussion that followed the reading of G. J. Lidstone's paper on Endowment Assurances, G. F. Hardy said that 'the errors in the successive groups formed a curve very similar to the normal curve of error' (*J. Inst. Actu.* XXXIV, 87), and the series in question is a rather interesting example of a symmetrical distribution.

Unexpired term in years	Error involved in using 'mean age' method
0–4	11
5–9	116
10–14	274
15–19	451
20–24	432
25–29	267
30–34	116
35, etc.	16
...	1683

Moments were calculated about the centre of the 15–19 group, and 0·4985146, 2·161022, 3·104576, and 12·60666 were found for the first four moments; transforming to the mean (17·5+2·492573 = 19·992573), and using Sheppard's adjustments, the following values result:

$$\mu_2 = 1{\cdot}829172 \qquad \mu_3 = 0{\cdot}120452 \qquad \mu_4 = 8{\cdot}52636$$
$$\beta_1 = 0{\cdot}0023706 \qquad \beta_2 = 2{\cdot}548313 \qquad \kappa = -0{\cdot}007492$$

which shows that Type II can be used.

The equations for the type give

$$m = 4{\cdot}141766, \qquad a = 4{\cdot}543079, \qquad y_0 = 462{\cdot}57$$

The mean and mode coincide, because the curve is symmetrical.

For calculating a series of values, the following arrangement is convenient:

$\frac{x}{a}$	$\log\left(1+\frac{x}{a}\right)$	$\log\left(1-\frac{x}{a}\right)$	(2)+(3)	$\log y_x = m\times(4) + \log y_0$
(1)	(2)	(3)	(4)	(5)

It is easier to work in this way than by calculating values of $1-x^2/a^2$. In the particular example, ordinates were calculated at the beginning, middle, and end of each group, and Simpson's quadrature formula was used for finding the areas, viz.

$$\int_0^1 y\,dx = \tfrac{1}{6}\{y_0+4y_{\frac{1}{2}}+y_1\}.$$

Group	Ungraduated figures	Areas Type II	Mid-ordinates, Type II	Areas, 'Normal curve'
0–4	11	14	11	22
5–9	116	109	104	95
10–14	274	286	287	270
15–19	451	433	440	455
20–24	432	433	440	455
25–29	267	285	287	269
30–34	116	109	104	95
35, etc.	16	14	11	22
...	1683	1683	...	1683

A comparison of the mid-ordinates with the areas gives an idea of the error involved in using the former for the latter; the differences are largest at the 'tails' and near the mode.

The curve starts at 19·992573—22·71540 = —2·72283, and ends at 42·70797.

The final column of the table gives a graduation by the 'normal curve'.

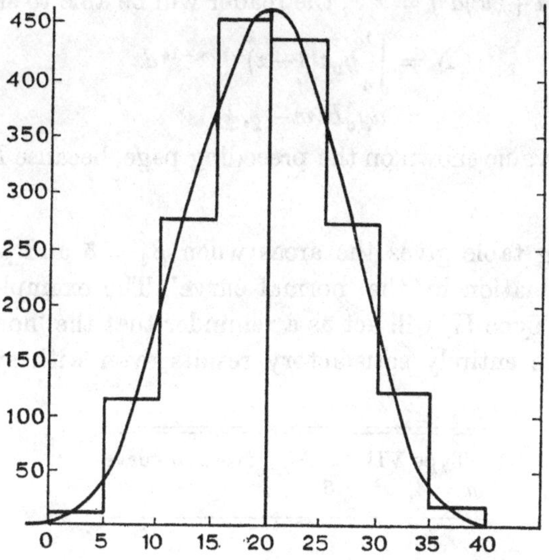

TYPE II

TRANSITION TYPE (TYPE VII)

$$y = y_0\left(1+\frac{x^2}{a^2}\right)^{-m}$$

Origin at mode (= mean)

$$m = \frac{5\beta_2-9}{2(\beta_2-3)}$$

$$a^2 = \frac{2\mu_2\beta_2}{\beta_2-3}$$

$$y_0 = \frac{N}{a\sqrt{\pi}}\cdot\frac{\Gamma(m)}{\Gamma(m-\frac{1}{2})}$$

Notes and Proof

The curve may be taken as a special case of Type IV when $\nu = 0$. This happens when $\beta_1 = 0$ and $\beta_2 > 3$. The curve is symmetrical and of unlimited range in both directions.

$$N = \int_{-\infty}^{+\infty} y_0\left(1+\frac{x^2}{a^2}\right)^{-m} dx$$

$$= 2\int_0^{\infty} y_0\left(1+\frac{x^2}{a^2}\right)_{-m} dx$$

then putting $1+(x^2/a^2) = z^{-1}$, the reader will be able to show that

$$N = \int_0^1 y_0 a(1-z)^{-\frac{1}{2}} z^{m-1\frac{1}{2}}\, dz$$
$$= a y_0 B(m-\tfrac{1}{2}, \tfrac{1}{2})$$

or y_0 has the value shown on the preceding page, because $\Gamma(\frac{1}{2}) = \sqrt{\pi}$.

Example

The following table gives the areas when $\beta_2 = 5$ and $\mu_2 = 1$ and shows a graduation by the 'normal curve'. The example, together with that of Type II, will act as a reminder that the 'normal curve' need not give entirely satisfactory results even with symmetrical distributions.

Type VII $m = 4,\ a^2 = 5$	Normal curve $\sigma = 1$
1	...
1	...
2	...
4	...
7	1
16	5
38	24
93	93
225	278
527	656
1106	1210
1858	1746
2244	1974
1858	1746
etc.	etc.

TRANSITION TYPE (TYPE III)

$$y = y_0\left(1+\frac{x}{a}\right)^{\gamma a} e^{-\gamma x} \qquad (x > -a)$$

Origin at mode

$$\gamma = \frac{2\mu_2}{\mu_3}$$

$$p = \gamma a = \frac{4}{\beta_1} - 1$$

$$a = \frac{2\mu_2^2}{\mu_3} - \frac{\mu_3}{2\mu_2}$$

$$y_0 = \frac{N}{a} \cdot \frac{p^{p+1}}{e^p \Gamma(p+1)}$$

$$\text{Mode} = \text{Mean} - \frac{\mu_3}{2\mu_2}$$

If expressing curve with origin at mean (see Table 6, on page 45):

$$y_e = N \, . \, \gamma \, . \, \frac{(p+1)^p}{e^{p+1}\Gamma(p+1)}$$

Notes

The curve is usually bell-shaped, but becomes J-shaped when $p < 0$, that is, when $\beta_1 > 4$. The range is limited in one direction only. The criterion is that $2\beta_2 = 6+3\beta_1$. Theoretically this gives $\kappa = \infty$ but the curve may be used in many cases where κ is not very large, provided $2\beta_2$ approximates to $6+3\beta_1$. If $\beta_1 < 4$, γ and a have the same sign for $0 < \sqrt{\beta_1} < 2$, γ and a are positive, so that the range is limited from below at a distance of a below the mode; when μ_3 is negative γ and a are negative, so that the range is limited from above at a distance a above the mode. If, however, $\beta_1 > 4$, then a and γ have different signs.

Example

The following statistics are taken from a paper in the *Trans. Actu. Soc. Edinb.* IV, 44, and give the numbers of wives tabulated for the ages of mothers, and according to years since marriage. The mothers' ages for the particular series are 30 to 34.

Years after marriage	Number of wives	Graduated by Type III curve
0–1	44	59
1–2	135	111
2–3	45	45
3–5	12	20
4–5	8	9
5–6	3	4
6–7	1	2
7–8	3	1
Total	251	251

The mean is 1·8346612 and the moments about the centroid vertical are 1·441787, 3·606622 and 18·93221; so that $\kappa = -8{\cdot}44$.

As this value was large, Type III was used, and

$$\gamma = 0{\cdot}7995221 \qquad a = -0{\cdot}098007$$
$$p = -0{\cdot}0783584 \qquad y_0 = 214{\cdot}8$$

This example is given because it can be used to show a difficulty rather clearly. At first sight, a curve starting at zero, rising to a maximum, and then falling, might be expected. Instead, we find the curve starting at duration 0·68192 *; so that the first group is

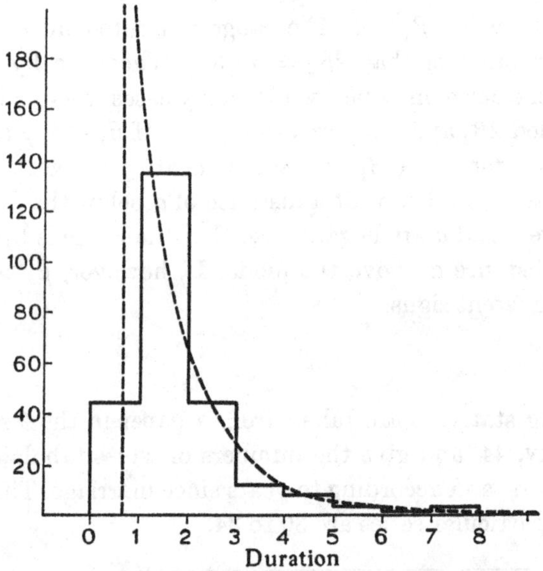

made up of a strip on a base 0·31808 in length, and has a smaller value than the next group, though any ordinate read off within the first group would be larger than any ordinate in the second group. No adjustment was made to the rough moments.

Proof

In the equation for the type, viz. $y = y_0(1+x/a)^{\gamma a}\, e^{-\gamma x}$, put $\gamma a = p$, and substitute z for $\gamma(a+x)$; then, if N be the total frequency,

* The mode in ordinary cases of Type III is given by mean $-\mu_3/2\mu_2$. In this case, $(\mu_3/2\mu_2) = 1{\cdot}25075$; so the mode would be at 0·53391, and the curve would start at {'mode' $-a$} $= 0{\cdot}58391 + 0{\cdot}09801 = 0{\cdot}68192$.

$$N = \int_{-a}^{\infty} y_0\left(1+\frac{x}{a}\right)^p e^{-\gamma x}\,dx$$

$$= \int_0^{\infty} y_0 z^p a^{-p}\, e^{-z+p}\gamma^{-(p+1)}\,dz \text{ for } \frac{dz}{dx} = \gamma$$

$$= y_0\frac{e^p}{\gamma p^p}\int_0^{\infty} z^p\, e^{-z}\,dz$$

$$= y_0\frac{ae^p}{p^{p+1}}\Gamma(p+1)$$

This gives $y_0 = \dfrac{Np^{p+1}}{ae^p\Gamma(p+1)}$

The nth moment about the start of the curve is

$$\frac{1}{N}\int_{-a}^{\infty} y_0\left(1+\frac{x}{a}\right)^p e^{-\gamma x}\,(x+a)^n dx = \frac{y_0 e^p}{Np^p\gamma^{n+1}}\int_0^{\infty} z^{p+n}\, e^{-z}\,dz$$

$$= \frac{\Gamma(p+n+1)}{\gamma^n\Gamma(p+1)}$$

by using the value of y_0 found above.

Since $\Gamma(p) = (p-1)\Gamma(p-1)$, the first moment is $(p+1)/\gamma$, the second $[(p+1)(p+2)]/\gamma^2$, and the third $[(p+1)(p+2)(p+3)]/\gamma^3$. In order to apply these formulae to statistical work, it is necessary to have moments about the centroid vertical, the position of which (the mean) can be found; and as, by definition, the first moment about it is zero, we get

$$\mu_2 = \frac{p+1}{\gamma^2} \quad\text{and}\quad \mu_3 = \frac{2(p+1)}{\gamma^3}$$

These results give γ and p as $2\mu_2/\mu_3$ and $(4\mu_2^3/\mu_3^2)-1$ respectively.

TRANSITION CURVE (TYPE V)

$y = y_0 x^{-p}\, e^{-\gamma/x} \qquad (x > 0) \qquad\qquad (p>1)$

Origin at start of curve

$p = 4 + \dfrac{8+4\sqrt{\{4+\beta_1\}}}{\beta_1} \qquad\qquad (p>4)$

$\gamma = (p-2)\sqrt{\{\mu_2(p-3)\}} \qquad\qquad (p>3)$

$y_0 = \dfrac{N\gamma^{p-1}}{\Gamma(p-1)}$

$$\text{Origin} = \text{Mean} - \frac{\gamma}{p-2}$$

$$\text{Mode} = \text{Mean} - \frac{2\gamma}{p(p-2)}$$

The sign of γ is the same as that of μ_3.

If expressing curve with origin at mean (see Table 6, on page 45):

$$A = \gamma/(p-2)$$

$$y_e = \frac{N(p-2)^p}{\gamma e^{p-2}\Gamma(p-1)}$$

Example

The following series of deaths is taken from G. King's paper 'On the rate of mortality amongst female nominees, etc.' (*J. Inst. Actu.* XXXIII, 262–8):

Ages	Deaths	Graduated by Type V
30–34	1	1
35–39	5	3
40–44	8	6
45–49	12	14
50–54	28	32
55–59	82	68
60–64	128	137
65–69	253	247
70–74	342	381
75–79	525	480
80–84	438	441
85–89	265	261
90–94	53	80
95–99	18	10
100, etc.	4	1
...	2162	2162

The mean is at age 75·9782605, and the moments (adjusted), etc. are (in 5-year units)

$\mu_2 = 3{\cdot}573346$ $\quad \mu_3 = -4{\cdot}752613$ $\quad \mu_4 = 51{\cdot}02583$

$\beta_1 = 0{\cdot}4950399$ $\quad \beta_2 = 3{\cdot}996134$ $\quad \kappa = 0{\cdot}85$

Strictly speaking, Type IV should be used, but the value of κ is not very far from unity, and the following Type V constants were found:

$$p = 37{\cdot}29145$$
$$\gamma = -390{\cdot}6609 \text{ (negative, because } \mu_3 \text{ is)}$$
$$\log y_0 = 56{\cdot}930518$$

The approximation to the value of $\log \Gamma(p-1)$ was used. The origin is at age 131·32606, and the mode at 78·9467.

The columns used for calculating the ordinates were:

x	$\log x$	$-p \log x$	$\frac{1}{x}(-\gamma \log_{10} e)$	$\log y = \log y_0 + (3) + (4)$	$y = \text{antilog } (5)$
(1)	(2)	(3)	(4)	(5)	(6)

The logarithms are *all* common logarithms (to base 10) although the base is only inserted in $\log_{10} e$. Col. (4) is most easily computed using a table of reciprocals, with the constant multiplier $(-\gamma \log_{10} e)$.

The point to be borne in mind in drawing a curve of this type is that as the mode and origin are not at the same place, care must be taken to give the maximum ordinate its right position and magnitude (cf. Type IV).

The graduated figures agree fairly closely with the original statistics below the 90–94 group, but are unsuitable for that and the two later groups. The reason is that Type IV, having an unlimited range, should be used. The particular case was chosen partly because an example in which μ_3 is negative is rather more awkward than when μ_3 is positive. In such cases it is a good check to imagine the statistics written in inverse order (in this case 4, 18, 53, etc.), and so avoid the negative signs.

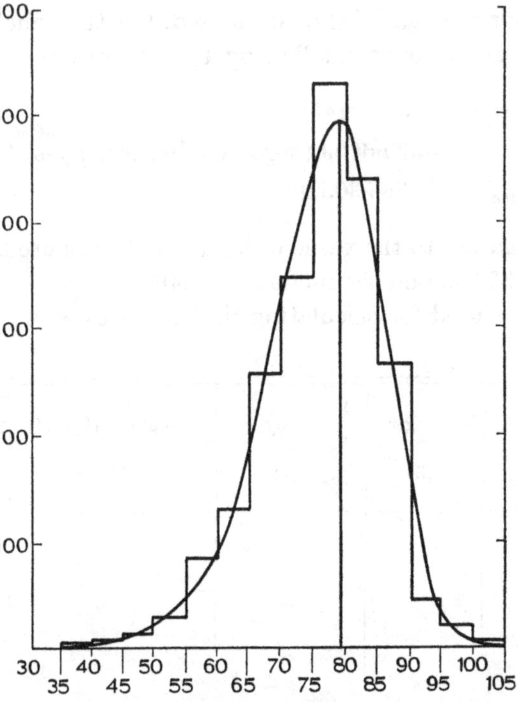

TYPE V

Proof

Putting $\gamma/x = z$ in $y = y_0 e^{-\gamma/x} x^{-p}$, and integrating from 0 to ∞, we have

$$N = y_0 \gamma^{1-p} \Gamma(p-1)$$

or
$$y_0 = \frac{N\gamma^{p-1}}{\Gamma(p-1)}$$

Using the same substitution, the nth moment about the origin is

$$\mu_n' = \frac{y_0}{N}\gamma^{n-p+1}\int_0^\infty e^{-z} z^{p-n-2}\,dz$$

$$= \frac{y_0}{N}\gamma^{n-p+1}\Gamma(p-n-1)$$

$$= \gamma^n \frac{\Gamma(p-n-1)}{\Gamma(p-1)} \qquad (n<p-1)$$

This gives
$$\mu_1' = \frac{\gamma}{p-2} \qquad (p>2)$$

which is the distance between the mean and origin,

$$\mu_2' = \frac{\gamma^2}{(p-2)(p-3)} \qquad (p>3)$$

$$\mu_3' = \frac{\gamma^3}{(p-2)(p-3)(p-4)} \qquad (p>4)$$

Transferring the moments to the centroid vertical,

$$\mu_2 = \frac{\gamma^2}{(p-2)^2(p-3)} \qquad (p>3)$$

and

$$\mu_3 = \frac{4\gamma^3}{(p-2)^3(p-3)(p-4)} \qquad (p>4)$$

$\therefore$

$$\beta_1 = \frac{\mu_3^2}{\mu_2^3} = \frac{16(p-3)}{(p-4)^2} = \frac{16}{p-4} + \frac{16}{(p-4)^2}$$

and

$$(p-4)^2 - \frac{16}{\beta_1}(p-4) - \frac{16}{\beta_1} = 0$$

$p-4$ will have to be taken as the positive root of the equation. Curves with $p<4$ are possible but cannot be obtained by this method.

Since the tangent to the curve at the top of the maximum ordinate is parallel to the axis of x, the position of the mode is such that dy/dx is zero there, i.e. $y_0 x^{-p-1} e^{-\gamma/x}\{-p+(\gamma/x)\}$ is zero. $x = 0$ and $x = \infty$ give the cases in which the curve touches the axis of x, and the other case, the one required, is when $p-(\gamma/x) = 0$, or $x = (\gamma/p)$, i.e. the mode is γ/p from the origin.

UNCOMMON FREQUENCY TYPES

9. Although we have now described all the possible types of Pearson curves, there are some cases of particular interest, which merit individual treatment.

We will now deal with these cases, but we shall discuss them in less detail than the more important types.

Type VIII

$$y = y_0\left(1+\frac{x}{a}\right)^{-m}$$

Range from an infinite ordinate at $-a$ to a finite ordinate y_0, at 0.

m is found from the solution of

$$m^3(4-\beta_1)+m^2(9\beta_1-12)-24\beta_1 m+16\beta_1 = 0$$

and must be neither * <0 nor >1.

$$a = \pm\sigma(2-m)\sqrt{\frac{3-m}{1-m}}$$

$$y_0 = N(1-m)/a$$

The mean is $-a+a(1-m)/(2-m) = -a/(2-m)$. When μ_3 is positive a is negative.

If we use the form with origin at mean (see Table 6, on page 45)

$$A = a(1-m)/(2-m) \quad \text{and} \quad y_e = N(1-m)(2-m)^m/a(1-m)^m.$$

The curve is a special case of Type I when m_2 is zero, that is, when

$$r-2 = r(r+2)\sqrt{\{\beta_1/[\beta_1(r+2)^2+16(r+1)]\}}$$

where $\qquad r = 6(\beta_2-\beta_1-1)/(6+3\beta_1-2\beta_2)$

Thus the criterion for the suitability of the curve is that

$$\frac{(4\beta_2-3\beta_1)(10\beta_2-12\beta_1-18)^2-\beta_1(\beta_2+3)^2(8\beta_2-9\beta_1-12)}{(3\beta_1-2\beta_2+6)\{\beta_1(\beta_2+3)^2+4(4\beta_2-3\beta_1)(3\beta_1-2\beta_2+6)\}}$$

or λ, say, is zero.

The criteria for Type VIII can be reduced to (1) special case of Type I, (2) $\lambda = 0$, (3) $5\beta_2-6\beta_1-9$ is negative. It may be added that $24\beta_2-27\beta_1-38$ is small; theoretically positive.

If $\beta_1 = 0$ an interesting special case arises, in which $m = 0$, and the curve becomes a horizontal line, which is also the limit of Types IX and XII. This is also a special Type II curve and is called the *uniform* or *rectangular* curve.

The solution of the cubic for m gives trouble. m can also be found from $m = -2(5\beta_2-6\beta_1-9)/(3\beta_1-2\beta_2+6)$, and though this involves β_2 it should theoretically give the same value of m as the cubic. As the criterion is not exactly reached in practice the two results differ, and it seems preferable to find m from the cubic by using

$$m = \frac{24\beta_1-\sqrt{\{24^2\beta_1^2-64\beta_1[m'(4-\beta_1)+9\beta_1-12]\}},}{2\{m'(4-\beta_1)+9\beta_1-12\}}$$

where m' is found from the expression in β_1 and β_2 given above or by some other trial method.

* The case $m>1$ is impossible; the case $m<0$ corresponds to Type IX.

An alternative is to find from the criteria or from the diagram in *Biometrika Tables* the value of β_2 which is the consequence of the particular value of β_1 when a Type VIII curve occurs, and use this theoretical value in finding m instead of the β_2 given by the actual statistics.

Example

Frequency	Graduation (1)	Graduation (2)
469	437	436
186	222	209
166	165	161
134	136	141
122	120	127
112	109	115
1189	1189	1189

The mean is 0·65518 of an interval after the centre of the 186 group. The constants were

$$\mu_2 = 2{\cdot}986 \qquad \mu_3 = 3{\cdot}295 \qquad \mu_4 = 18{\cdot}252$$
$$\beta_1 = 0{\cdot}408 \qquad \beta_2 = 2{\cdot}047 \qquad \lambda = -0{\cdot}05$$

$5\beta_2 - 6\beta_1 - 9$ negative. Hence Type VIII can be used.

$$m = 0{\cdot}500$$
$$a = -5{\cdot}797$$
$$y_0 = 102{\cdot}6$$

The curve runs from 0·277 before the middle of the first to 0·02 after the end of the last group. The graduation is shown (no. 1) above.

The areas can be calculated by the expression

$$y_{-r}(a-r)/(1-m)$$

which gives the area of the remainder from $-r$ to $-a$. In the particular case the range could be fixed at 6 as the data related to six months' experience of maturities among endowment assurances, and remembering that the mean is

$$a(1-m)/(2-m)$$

we found

$$m = 0{\cdot}439 \qquad y_0 = 111{\cdot}1$$

The areas resulting are given in graduation (2). The following table gives the calculation of the areas in this case. The equation of the curve is

$$y = 111{\cdot}1\left(1-\frac{x}{6}\right)^{-0{\cdot}439}$$

with range from 0 to 6, a being negative because μ_3 is positive.

x	$1-x/6$	Colog (2)	(3) × m	(4) + log 111·1 = log y_x	log $\frac{y_x}{1-m}$	Antilog (6)	Remainder of range	(7) × (8)	Area required
(1)	(2)	(3)	(4)	(5)	(6)	(7)	(8)	(9)	(10)
0	...	...	...	...	...	...	6	1189	115
1	0·8333	0·0792	0·0348	2·0804	2·3318	214·7	5	1074	127
2	0·6667	0·1761	0·0774	2·1230	2·3744	236·8	4	947	141
3	0·5000	0·3010	0·1323	2·1779	2·4293	268·7	3	806	161
4	0·3333	0·4815	0·2118	2·2574	2·5088	322·7	2	645	209
5	0·1667	0·7781	0·3419	2·3875	2·6389	435·5	1	436	436

Some of the columns can be dispensed with: they are shown in detail to make the method clear.

Both graduations are reasonably close to the facts.

An example of the limiting case will be found in the following statistics:

No.	Frequency	Graduation	Theoretical
1	45	42	45
2	38	45	45
3	46	45	45
4	53	45	45
5	43	45	45
6	38	45	45
7	49	45	45
8	41	45	45
9	44	45	45
10	52	48	45
...	449	450	450

The mean is 0·57 after the middle of the 5 group; the moments are

$$\mu_2 = 8{\cdot}374 \qquad \mu_3 = 0{\cdot}026 \qquad \mu_4 = 124{\cdot}46$$
$$\beta_1 = 0{\cdot}011 \qquad \beta_2 = 1{\cdot}78$$

We take $m = 0$

Hence
$$y = \frac{449}{2\sqrt{(3\mu_2)}} = 44{\cdot}8$$

The range is from 0·57 to 10·43.

The series was found by summing in tens the last figure of Carlisle $3\frac{1}{2}$ per cent. Table of A_x. The range should be 0·5 to 10·5. On the assumption of a uniform distribution, one would expect the mean to be 5·5 instead of 5·7, and y would be 45. The example is interesting as showing how the Pearson-curves graduate in an extreme case. The 'graduation' and theoretical results are shown. In the 'graduation' decimals have been neglected.

Type IX

$$y = y_0\left(1+\frac{x}{a}\right)^m$$

Range from $x = -a$ where $y = 0$ to $x = 0$ where $y = y_0$

$$a = \pm\sigma(m+2)\sqrt{\left(\frac{m+3}{m+1}\right)}$$

m is found by solving

$$m^3(\beta_1-4)+m^2(9\beta_1-12)+24m\beta_1+16\beta_1 = 0$$

$$y = N(m+1)/a$$

Note that m must not be negative.

The distance of the mean from $x = -a$ is $a(m+1)/(m+2)$ and from $x = 0$ is $-a/(m+2)$.

If we use the form with origin at mean (see Table 6, on page 45.)

$$A = \frac{(m+1)a}{(m+2)} \quad \text{and} \quad y_e = \frac{N(m+1)^{m+1}}{a(m+2)^m}$$

As in Type VIII, the value of m can be found by simplifying the cubic into a quadratic, or by the other method indicated.

The criteria are reached through the same equation as those for Type VIII, and can be reduced to (1) special case of Type I, (2) $\lambda = 0$, (3) $5\beta_2-6\beta_1-9$ is positive, (4) $2\beta_2-3\beta_1-6$ is negative.

If $\beta_2 = 2{\cdot}4$ and $\beta_1 = 0{\cdot}32$, the curve becomes a sloping line

$$y = \frac{\sqrt{(2)}N}{3\sigma}\left(1+\frac{x}{3\sqrt{(2)}\sigma}\right)$$

This is sometimes called a *trapezoidal* curve.

If $\beta_1 = 0$ we reach a horizontal line as the limit, while if $\beta_2 = 9$ and $\beta_1 = 4$, we have the other limit of Type IX, and find the exponential series (Type X).

7

Example

The mean is at 2·909 assuming the exposed to risk to be an ordinate at the duration or an area from $n-\frac{1}{2}$ to $n+\frac{1}{2}$.

$$\mu_2 = 6{\cdot}27 \qquad \mu_3 = 10{\cdot}99 \qquad \mu_4 = 102{\cdot}50$$
$$\beta_1 = 0{\cdot}490 \qquad \beta_2 = 2{\cdot}606$$
$$5\beta_2-6\beta_1-9 \text{ is positive} \qquad 2\beta_2-3\beta_1-6 \text{ is negative}$$

Duration	Exposed to risk in annuity experience	Type IX	Frequency line
0	119	118	108
1	100	98	97
2	86	85	86
3	75	74	76
4	61	63	65
5	50	52	54
6	39	41	43
7	27	30	32
8	22	20	22
9	12	11	11
10	3	2	0
...	594	594	594

The curve is not far from Type IX, and if β_1 had been 0·32 and β_2 had been 2·4, we should have reached a straight line

$$y = 111{\cdot}8\left(1-\frac{x}{10{\cdot}63}\right)$$

with range from $-0{\cdot}6$ to 10·0 and obtained the graduation shown. The whole area $-0{\cdot}6$ to $+0{\cdot}5$ is taken as the frequency for duration 0. Using Type IX, the following constants are reached:

$$m = 1{\cdot}123, \qquad a = -10{\cdot}913, \qquad y_0 = 115{\cdot}54$$

The curve runs from $-0{\cdot}586$ to 10·3275. The 118 in the first group has been taken as the area from $-0{\cdot}586$ to $+0{\cdot}5$. Theoretically there cannot be an exposure before duration $-0{\cdot}5$, but as we are merely giving an example of fitting a curve to a series of numbers this need not concern us. The difficulty could be met by fitting a system of ordinates or by assuming a starting point for the curve.

If m happens to be less than unity the shape of the curve is somewhat different, e.g. if $y = 100(1+x/10)^{0{\cdot}25}$ we have the following ordinates:

100, 98, 95, 91, 88, 84, 79, 74, 67, 56, 0

The actual deaths in a select mortality experience may take this form, but the shape of the curve will be less flat at the start, e.g. in the American Medico-Actuarial experience 1913 age group 30–34.

Type X

$$y = \frac{N}{\sigma} e^{-x/\sigma}$$

Range from 0 to ∞.

Distance of origin from the mean is σ.

The ordinate at the mean (y_e) is $N/e\sigma$.

(If μ_3 is negative, $e^{-x/\sigma}$ is replaced by $e^{x/\sigma}$ and the range is from $-\infty$ to zero.)

The curve is a special case of Type III when $\gamma a = 0$, that is when $\beta_1 = 4$.

The condition for Type III is given by $2\beta_2 = 6+3\beta_1$. Hence the exponential form is given by $\beta_1 = 4$ and $\beta_2 = 9$. The curve is also the limit of Types IX and XI.

This is called an exponential curve. In the following example the 'theoretical' figures correspond to $\sigma = (\log_e 2)^{-1}$, $N = 8192$. The 'data' were obtained as a result of coin-tossing experiments.

Example

Values of x	Frequency	Graduation	Theoretical
0–1	4165	4132	4096
1–2	2028	2016	2048
2–3	982	1015	1024
3–4	480	511	512
4–5	266	257	256
5–6	132	130	128
6–7	71	65	64
7–8	36	33	32
8–9	17	17	16
9–10	9	8	8
10–11	2	4	4
11–12	1	2	2
12–13	1	1	1
13–14	1	1	1
>14	1	...	...
...	8192	8192	8192

The observed mean is 2·0087.

$$\mu_2 = 2{\cdot}045 \qquad \mu_3 = 6{\cdot}290 \qquad \mu_4 = 39{\cdot}720$$
$$\beta_1 = 4{\cdot}629 \qquad \beta_2 = 9{\cdot}502 \qquad \sigma = 1{\cdot}43$$

When the curve is an exponential the moments and mean require adjustment, but the Sheppard high contact adjustments are, of course, unsuitable. If the curve starts at the beginning of the first group, the mean is overstated on average when μ_3 is positive by $\frac{1}{2}+(e^{1/\sigma}-1)^{-1}-\sigma$, that is, $\frac{1}{12}\sigma^{-1}$ approximately, if σ is large and the second moment about the true mean is understated by $\frac{1}{12}$ approximately.* Making use of the adjustments the mean is now 1·934, μ_2 is 2·123, and σ is 1·457.

It is interesting to notice how the 'graduation' keeps closer to the frequency than the theoretical result.

I give as a second example the following series based on cricket scores, known to start at the beginning of the first group:

Score	0–19	20–	40–	60–	80–	100–	120–	140–	160–
Series	64	34	18	9	6	3	3	0	0
Graduation	64	34	18	10	5	3	1	1	1

The ratio of each term in the graduated figures to the preceding is 0·54, and the graduation is almost exact. Owing, however, to the 3 at the group 120, the moments give a criterion considerably removed from the theoretical $\beta_1 = 4$, $\beta_2 = 9$.

Type XI

$$y = y_0 x^{-m}$$

Range from $x = b\ (>0)$ where $y = y_0 b^{-m}$ to $x = \infty$ where $y = 0$.
m is found from

$$m^3(4-\beta_1)+m^2(9\beta_1-12)-24\beta_1 m+16\beta_1 = 0$$

$$b = \pm\sigma(m-2)\sqrt{\frac{m-3}{m-1}}$$

$$y_0 = Nb^{m-1}(m-1)$$

The distance of mean from origin is $b(m-1)/(m-2)$.

* See the discussion in Appendix 1.

If μ_3 is negative, $y = y_0|x|^{-m}$, and the range is from $-\infty$ to $-b(b > 0)$.

If we use the form with origin at mean (see Table 6, on page 45), $A = b(m-1)/(m-2)$ and

$$y_e = \frac{N}{b} \cdot \frac{(m-2)^m}{(m-1)^{m-1}}$$

As in Type VIII, m can be found by simplifying the cubic into a quadratic or by the other method indicated.

m must be greater than 1, and is usually considerably greater than 1.

The curve is a special case of Type VI when $q_2 = 0$.

The criteria can be expressed as (1) special case of Type VI, (2) $\lambda = 0$, (3) $2\beta_2 - 3\beta_1 - 6$ is positive.

Example

Duration	Withdrawals	Graduation by XI
0	165	183
1	65	53·9
2	23	32·6
3	32	20·0
4	13	12·4
5	8	7·6
6	1	4·9
7	6	3·1
8	3	1·9
9	3	1·2
10	1	0·8
11	3	1·6
...	323	323

I have not come across a distribution really represented by this type, but I give an unsuccessful attempt to apply it to a series of withdrawals. The constants were

$$\beta_1 = 4{\cdot}97 \qquad b = 57{\cdot}14$$

$$m = 29{\cdot}69 \qquad \log y_0 = 54{\cdot}3563$$

Distance of mean from origin = 59·205

In calculating areas we use $y_0 a^{-(m-1)}/(m-1)$ as the area from a to ∞.

Twisted J-shaped Curve

As pointed out in the notes on Type I (p. 52), we obtain an interesting curve when both m_1 and m_2 are numerically less than unity and one of them is negative. It arises when

$$\beta_2 > 1{\cdot}5+1{\cdot}125\beta_1$$

and when $$\beta_2 < 2+1{\cdot}25\beta_1$$

as can be seen by remembering that the sum of the values of the m's must lie between 1 and -1 or r lies between 3 and 1. A special case has been discussed as a transition type (No. XII) when

$$y = y_0\left(\frac{\sigma\{\sqrt{(3+\beta_1)}+\sqrt{\beta_1}\}+x}{\sigma\{\sqrt{(3+\beta_1)}-\sqrt{\beta_1}\}-x}\right)^{\sqrt{\{\beta_1/(3+\beta_1)\}}}$$

Range from $x = \sigma(\sqrt{(3+\beta_1)}-\sqrt{\beta_1})$ to $x = -\sigma(\sqrt{(3+\beta_1)}+\sqrt{\beta_1})$. The origin is at the mean.

$$y_0 = \frac{N}{b\Gamma(m+1)\Gamma(1-m)}$$

where $m = \sqrt{[\beta_1/(3+\beta_1)]}$ and $b = 2\sigma\sqrt{(3+\beta_1)}$.

When μ_3 is positive, the negative sign is taken for the square roots.

The limit of the curve when $\beta_1 = 0$ is a horizontal line.

The criterion is $5\beta_2-6\beta_1-9 = 0$.

Example

Frequency	Graduation	Ordinates
...	2	18·5
33	31	31·6
...	...	40·6
53	49	49·3
...	...	57·0
65	65	65·2
...	...	73·4
81	83	82·4
...	...	92·3
101	103	103·3
...	...	114·6
131	134	132·9
...	...	155·5
186	191	186·0
...	...	244·2
350	342	405·4
1000	1000	...

The mean is 0·051 after the centre of the 131 group. The constants are

$$\mu_2 = 4{\cdot}266 \qquad \mu_3 = -7{\cdot}688 \qquad \mu_4 = 48{\cdot}154$$

$$\beta_1 = 0{\cdot}761 \qquad \beta_2 = 2{\cdot}646 \qquad 5\beta_2 - 6\beta_1 - 9 = -0{\cdot}368$$

$$y = 87{\cdot}2\{(5{\cdot}808+x)/(2{\cdot}204-x)\}^{0{\cdot}45}$$

In addition to the graduation a number of equidistant ordinates is given. They show that the curve rises abruptly, then less abruptly and then again more abruptly. The withdrawals in select tables are sometimes of this shape (e.g. Japanese experience, 1910, age 52, females). A somewhat similar twist occurs in a population curve.

U-*shaped Curve*

This shape arises in Type I when m_1 and m_2 (or m in Type II) are negative. There are difficulties in fitting it to statistics because it is awkward to adjust the rough moments. Abruptness corrections are needed at both ends of the distribution, and are not easy to assess.

The limit of the U-curve is two separate blocks of frequency at the ends of the range. This limit is reached when

$$\beta_2 - \beta_1 - 1 = 0.$$

Some of the curves with which we have dealt are rare and in practical curve-fitting may be avoided, for they depend on certain definite values of β_1 and β_2, and the chance of reaching these exact values is negligible. In other words, if the object of fitting a curve in any particular case is to obtain the closest agreement between the actual figures given and the graduated figures, then the main Types (I, IV and VI) are all that are necessary, for the other types being transition types and depending on specific values of β_1 and β_2 need not arise. If, however, our object is to study probability in a wider sense, the transition types are of importance and they may, of course, be properly used when the values of the β's only differ from those indicated by the criteria to a small extent. This 'small extent' means within limits suggested by the standard errors of the β's.

10. As a result of an investigation by E. S. Pearson and J. W. Tukey,* some useful approximate formulae for Pearson frequency curves are available. It is convenient to introduce a notation for

* *Biometrika*, LII (1965), 533–46. (This paper also includes some interesting comparisons between Pearson curves and those to be described in Chapter 7.)

the percentiles of the curves. The θ percentile will be denoted by $[\theta\%]$ and is such that

$$\int_{-\infty}^{[\theta\%]} f_x dx = (\theta/100)\int_{-\infty}^{\infty} f_x dx$$

Then

$$\text{Mean} \doteqdot [50\%]+0{\cdot}185\Delta$$

where $\qquad \Delta = [95\%]+[5\%]-2[50\%].$

($[50\%]$ is, of course, the median.)

The suggested estimate of standard deviation is the larger of

$$\sigma' = \frac{[95\%]-[5\%]}{\max\{3{\cdot}29-0{\cdot}1(\Delta/\sigma')^2,\ 3{\cdot}08\}}$$

and

$$\sigma'' = \frac{[97{\cdot}5\%]-[2{\cdot}5\%]}{\max\{3{\cdot}98-0{\cdot}138(\Delta/\sigma'')^2,\ 3{\cdot}66\}}$$

As first approximations, it is suggested that

$$\sigma_1' = \frac{[95\%]-[5\%]}{3{\cdot}25} \quad \text{or} \quad \sigma_1'' = \frac{[97{\cdot}5\%]-[2{\cdot}5\%]}{3{\cdot}98}$$

(or an average of these) might be used.

As an example, consider a χ^2 distribution with 4 degrees of freedom. For this distribution

$$[2{\cdot}5\%] = 0{\cdot}4844; \qquad [5\%] = 0{\cdot}7107; \qquad [50\%] = 3{\cdot}3567$$
$$[95\%] = 9{\cdot}4877; \qquad [97{\cdot}5\%] = 11{\cdot}1433$$

and so $\Delta = 3{\cdot}4850$, and the estimate of the mean is $3{\cdot}3567+0{\cdot}6447 = 4{\cdot}001$

$$\sigma_1' = \frac{8{\cdot}7770}{3{\cdot}25} = 2{\cdot}701; \qquad \sigma_1'' = \frac{10{\cdot}6589}{3{\cdot}98} = 2{\cdot}678$$

Taking 2·69 as a first trial value we find

$$(\Delta/2{\cdot}69)^2 = 1{\cdot}678$$

Then

(i) $\quad 3{\cdot}29-0{\cdot}1\times 1{\cdot}678 = 3{\cdot}12 > 3{\cdot}08,$

and so

$$\sigma' = \frac{8{\cdot}7770}{3{\cdot}12} = 2{\cdot}813;$$

(ii) $\quad 3{\cdot}98-0{\cdot}138\times 1{\cdot}678 = 3{\cdot}75 > 3{\cdot}66$

and so

$$\sigma'' = \frac{10{\cdot}6589}{3{\cdot}75} = 2{\cdot}842.$$

The larger value, 2·842, is then taken as the estimator, giving $(\Delta/2{\cdot}842)^2 = 1{\cdot}5037$. After one further iteration we obtain, as the estimate of standard deviation, max(2·795, 2·825) = 2·825. In fact, the mean is 4·000, and the standard deviation is $\sqrt{8} = 2{\cdot}828$.

Although this approximation was constructed especially for use with Pearson curves (and χ^2 has a Type III frequency curve) it can also be used with some confidence for many other distributions.

11. There are tables (Johnson *et al.*, *Biometrika*, I (1963), 459–98) which can be used to aid in calculations concerned with fitting Pearson curves (particularly Type IV). They give, for specified pairs of values of $\sqrt{\beta_1}$ and β_2, values of the standardized deviates corresponding to a number of percentile points. Thus, in the notation of the preceding paragraph, they give values of

$$([\theta\%]-\text{mean})/(\text{standard deviation})$$

for $\theta = 0{\cdot}25, 0{\cdot}5, 1, 2{\cdot}5, 5, 10, 25, 50, 75, 90, 95, 97{\cdot}5, 99, 99{\cdot}5, 99{\cdot}75$), for $\sqrt{\beta_1} = 0{\cdot}0\ (0{\cdot}1)\ 2{\cdot}0$ and various values of β_2. The values of β_2 given do not cover fairly a narrow region close to the line $\beta_2-\beta_1-1=0$; nor do they include values of β_2 greater than 9 for $\sqrt{\beta_1}$ less than or equal to 1·2. For $\sqrt{\beta_1}$ greater than 1·2, β_2 extends up to about $9+6(\sqrt{\beta_1}-1{\cdot}2)$.

As an example of the use of these tables, suppose it is desired to calculate the upper quartile of the curve fitted to the data in the table on page 60. Here we have

$$\sqrt{\beta_1} = -0{\cdot}07325; \qquad \sqrt{\beta_2} = 3{\cdot}1729$$

and we take $[\theta\%] = [75\%]$. From the tables, the following standardized deviates are obtained:

$\sqrt{\beta_1} = 0{\cdot}0$	$\sqrt{\beta_2} = 3{\cdot}0 : 0{\cdot}6745$
$0{\cdot}0$	$3{\cdot}2 : 0{\cdot}661$
$0{\cdot}1$	$3{\cdot}0 : 0{\cdot}6844$
$0{\cdot}1$	$3{\cdot}2 : 0{\cdot}671$

(Note that we look up the value of [25%] in the tables and reverse the sign, because $\sqrt{\beta_1}$ is negative). Linear interpolation gives 0·673. So we calculate the upper quartile as mean+0·673 (standard deviation) $= 44{\cdot}5772+0{\cdot}673\times10{\cdot}6391 = 51{\cdot}74$.

ADDITIONAL EXAMPLES

1. Up to the present we have merely considered examples with a view to illustrating the various types of frequency curves, but it seems advisable to consider some practical examples which may help to show the range of applicability of the curves in actuarial work, and give an opportunity of noticing a few difficulties which may arise in applying them.

The function with which actuaries generally wish to deal in practical work is not an exposed to risk or series of deaths or withdrawals, but the ratio between the deaths and the exposed; that is, with the rates of mortality, sickness, marriage, and withdrawal. An actuary studying frequency curves may therefore naturally ask whether any of these rates can be graduated by means of the curves we have examined, and, if they fail, must they be put aside for some other method? Now the first point to be considered is whether these rates are frequency distributions; if they are not, the use of the frequency curve is empirical. A rate of mortality gives the proportion of people at each age who die, and if we imagine 1000 persons exposed to risk at each integral age, the number of deaths would be 1000 times the rate of mortality, and this seems to show that it is possible to consider the rate of mortality as a distribution, though it is one that could hardly arise in actual experience. It is impossible to describe the rates of mortality or sickness by a single frequency curve. On the other hand, the rates of marriage are certainly much like frequency curves, and the rates of withdrawal, whether regarded according to age or duration, might take a form like our example in Type III. There are, however, practical objections to the direct operation on rates, even apart from the very exaggerated idea of frequency distributions in which it is necessary to indulge. The numbers exposed to risk at the end of any table become small, and a single death or marriage there gives a very large rate, while at several ages near there may be a zero rate shown by the ungraduated data. This is extremely awkward, as it tends to make the ratios dealt with far rougher in application than the actual observations are in fact, and we are forced to group the material before using it, which introduces an arbitrary practice which it is well to avoid so far as possible. It must not, of course, be inferred that a small number of say fifty or one hundred deaths must necessarily be grouped according to each year of age, but that even if there are two or three thousand the roughnesses introduced by the use of rates influence the result considerably. Graduating rates means that an equal weight is given to each rate of mortality which is far from the weight indicated by the exposed to risk.

2. It will be useful to consider a case bearing out these objections and then deal with a practical method of overcoming them. The statistics to be considered have been taken from a paper by M. Mackenzie Lees 'On Rates of Mortality and Marriage among daughters of Peers and Heirs Apparent, etc.' (*Trans. Fac. Actu.* I, 276), and may be summarized as on page 101.

The moments were calculated by the Summation Method, and were found, about the mean 28·77191, to be

$$\mu_2 = 63{\cdot}2092 \qquad \mu_3 = 627{\cdot}101 \qquad \mu_4 = 19\,103{\cdot}3$$
$$\beta_1 = 1{\cdot}557153 \qquad \beta_2 = 4{\cdot}781321$$

The criterion was $\kappa = -1{\cdot}5$, but as I had neglected the rate 0·0089 at 71 in calculating the moments, I used Type III. The inclusion of the rate at that age would have lengthened the curve and considerably increased the arithmetical value of the criterion. Moreover $2\beta_2$ approximates to $6+3\beta_1$.

The constants for Type III were

$$\gamma = 0{\cdot}201592 \qquad a = 7{\cdot}78189$$
$$p = 1{\cdot}56881 \qquad \text{Mode} = 23{\cdot}81128$$

The curve starts, therefore, at age 16·02939.

$$y_0 = 890{\cdot}05.$$

The rates resulting from this graduation are given in the table, and while they tend to show that the distribution of rates of marriage is closely allied to a frequency curve, they do not give a satisfactory graduation, and the failure is due almost entirely to the objections mentioned above. If we were examining the algebraic form taken by rates of marriage, we should begin by work on population data where the roughness of material is avoided by the large numbers of individuals dealt with; as, however, we are seeking for a graduation, we must see how these objections, which of course apply to some extent to any method of graduation, can be overcome. It has been remarked that the cause of the difficulty is that incorrect weights are given to the items used, and the most obvious suggestion is that the actual exposed and marriages should be graduated separately. This, however, entails a large amount of additional work and seems to overlook the fact that deviations in the exposed to risk and the marriages are not independent. A shorter method can be used which avoids both the double graduation and the error just indicated. This method consists of using a series allied to the exposed, and treating it as a hypothetical exposed to risk from which a new series of marriages can be calculated. The advantages are that we have only to make one graduation, and the weights of the various parts of the table are given approximately. In a similar way q_x can be graduated, and in this connection it may be remarked that as the exposed to risk is generally capable of being represented by a frequency-curve, it is natural to suggest that the hypothetical exposed might be taken as the simplest form assumed by such curves (normal curve); this is also convenient because the ordinates for such curves have been tabulated.

3. The hypothetical exposed can be fixed by trial or from the values of the exposed. The column E'_x in the table given on page 101 is taken from Sheppard's Tables in *Biometrika Tables*, x being taken as 3·06, 3·084, 3·108, 3·132, etc., and the entries were multiplied by 10^6. $M'_x = E'_x \times m_x$ was then formed and graduated. The following values were obtained for the M'_x series:

Mean $= 24{\cdot}85779$

$\mu_2 = 29{\cdot}5006 \qquad \mu_3 = 190{\cdot}112 \qquad \mu_4 = 4361{\cdot}12$

$\beta_1 = 1{\cdot}40775 \qquad \beta_2 = 5{\cdot}01114 \qquad \kappa = -7{\cdot}102$

As κ is large, Type III was used, and

$\gamma = 0{\cdot}310350 \qquad p = 1{\cdot}841405 \qquad a = 5{\cdot}933325$

$y_0 = 192{\cdot}625 \qquad \text{Mode} = 21{\cdot}63562$

The curve was then worked out and the rates of marriage in the final column were obtained by dividing M' by E'. They agree closely with the ungraduated figures.

4. A numerical example of the application of the method to the $\text{O}^{\text{NM(5)}}$ Table may now be given. The normal curve with $\sigma = 10$ and origin at age $52\frac{1}{2}$ was used, and the values were multiplied by q_x with the help of Crelle's Tables.

A part of the work was

$q \times E \times 10^5$	Age	Ordinate from Sheppard's Tables $= E$	Age	$q \times E \times 10^5$
810	52	0·3984439	53	801
597	51	0·3944793	54	850
644	50	0·3866681	55	875

Age x	Exposed to risk E	No. of marriages M_x	Rate of marriage m_x	Rate of marriage graduated by frequency curve	Hypothetical exposed E_x	No. of marriages M'_x	Rate of marriage graduated
15	3658·0	3	0·0008	...	3695·0	3	...
16	3603·0	8	0·0022	0·0027	3433·0	7	0·0018
17	3528·5	49	0·0139	0·0132	3187·0	44	0·0157
18	3393·5	114	0·0336	0·0332	2957·0	99	0·0350
19	3187·0	176	0·0552	0·0517	2742·0	151	0·0541
20	2945·0	219	0·0744	0·0667	2541·0	189	0·0695
21	2688·5	192	0·0714	0·0776	2354·0	168	0·0809
22	2443·0	211	0·0864	0·0846	2179·0	188	0·0880
23	2187·0	212	0·0969	0·0881	2016·0	196	0·0917
24	1956·0	194	0·0992	0·0889	1864·0	185	0·0920
25	1758·0	146	0·0831	0·0875	1723·0	143	0·0901
26	1583·5	137	0·0865	0·0845	1591·0	138	0·0861
27	1417·0	121	0·0854	0·0803	1469·0	126	0·0812
28	1270·5	105	0·0826	0·0753	1355·0	112	0·0754
29	1148·5	75	0·0653	0·0698	1249·0	82	0·0693
30	1068·0	60	0·0562	0·0640	1151·0	65	0·0631
31	984·0	64	0·0650	0·0583	1061·0	69	0·0569
32	904·5	41	0·0453	0·0528	976·0	43	0·0508
33	848·5	30	0·0354	0·0475	897·0	32	0·0452
34	802·0	39	0·0486	0·0425	825·0	40	0·0400
35	752·0	20	0·0266	0·0378	758·0	20	0·0352
36	711·0	25	0·0352	0·0335	696·0	25	0·0309
37	672·5	18	0·0268	0·0295	639·0	17	0·0270
38	638·0	11	0·0172	0·0260	586·0	10	0·0235
39	612·5	14	0·0229	0·0228	537·0	12	0·0205
40	586·5	15	0·0256	0·0199	492·0	12	0·0176
41	568·5	9	0·0158	0·0173	451·0	7	0·0151
42	541·5	6	0·0111	0·0151	412·0	5	0·0130
43	515·0	8	0·0155	0·0131	376·0	6	0·0112
44	491·5	2	0·0041	0·0113	345·0	1	0·0096
45	476·0	5	0·0105	0·0097	315·0	3	0·0082
46	454·0	5	0·0110	0·0084	288·0	3	0·0070
47	440·5	2	0·0045	0·0072	262·0	1	0·0060
48	416·0	5	0·0120	0·0062	239·0	3	0·0051
49	395·0	2	0·0051	0·0054	218·0	1	0·0044
50	378·5	...	...	0·0046	199·0	...	0·0037
51	363·5	...	...	0·0039	181·0	...	0·0031
52	348·5	1	0·0029	0·0034	165·0	...	0·0026
53	335·5	2	0·0089	0·0029	150·0	1	0·0022
54	317·5	...	...	0·0024	139·0	...	0·0019
55	304·0	...	...	0·0020	124·0	...	0·0016
56	291·0	3	0·0103	0·0018	...	1	...
57	278·5	1	0·0036	0·0015	...	...	...
58	261·0	...	...	0·0013	...	...	...
59	248·5	...	...	0·0011	...	...	...
60	234·5	...	...	0·0009	75·7	...	0·0007
61	219·5	1	0·0046	0·0007	...	...	...
62	209·5	...	...	0·0006	...	...	...
63	201·5	...	...	0·0005	...	...	...
64	191·0	...	...	0·0004	...	...	...
65	177·0	...	...	0·0004	45·7	...	0·0003
66	165·5	...	...	0·0003	...	...	...
67	154·0	...	...	0·0002	...	...	...
68	147·5	...	...	0·0002	...	...	...
69	135·5	...	...	0·0001	...	...	...
70	124·5	...	...	0·0001	27·2	...	0·0001
71	112·5	1	0·0089	...	...	...	...
72	105·5	...	...	...	...	...	...
73	95·0	...	...	...	...	...	...
74	84·5	...	...	...	...	...	...
75	79·0	...	...	...	...	...	...

Summing these entries ($q \times E \times 10^5$) in fives, I formed the following:

Age	$q \times E \times 10^5$
20	13
25	70
30	218
35	594
40	1 394
45	2 460
50	3 702
55	4 519
60	4 385
65	3 602
70	2 249
75	1 197
80	461
85	133
90	31
95	5
100	1
	25 034

The abbreviations (use of Crelle's Tables and grouping) were adopted to save labour, and as the figures were required for an example they are sufficiently accurate.

The following values were then found:

Mean age = 59·439762

$\mu_2 = 4{\cdot}584327$ $\quad \mu_3 = -0{\cdot}4999871$ $\quad \mu_4 = 61{\cdot}17014$

Type of curve.—No. I

$m_1 = 32{\cdot}81166$

$m_2 = 26{\cdot}57123$

$a_1 = 18{\cdot}78553$

$a_2 = 15{\cdot}21272$

$y_0 = 4609{\cdot}884$

Mode age = 59·730789

(The unit is 5 years of age.)

The ordinates were then calculated for every fifth age, and finding that the curve is not very far removed from the normal curve of error, I interpolated in the second differences of the logarithms of the ordinates for those at the other ages.* A quadrature formula

* The logarithm of the normal curve is a quadratic function of the variable.

was used for finding areas, and q_x was found by dividing by the hypothetical figure already used for the exposed.

The expected deaths were as follows:

Group	Graduated q_x for central age of group	Actual	Expected	Deviation +	Deviation −
15–19	...	...	1·5	1·5	...
20–	0·00643	9	8·9	...	0·1
25–	0·00731	69	61·0	...	8·0
30–	0·00850	205	204·6	...	0·4
35–	0·00991	369	380·7	11·7	...
40–	0·01179	588	575·6	...	12·4
45–	0·01452	801	811·4	10·4	...
50–	0·01866	1 064	1 063·8	...	0·2
55–	0·02505	1 399	1 386·6	...	12·4
60–	0·03516	1 752	1 773·2	21·2	...
65–	0·05118	2 164	2 136·7	...	27·3
70–	0·07682	2 216	2 261·2	45·2	...
75–	0·11648	1 965	1 925·8	...	39·2
80–	0·17462	1 237	1 241·9	4·9	...
85–	0·24870	494	514·4	20·4	...
90–	0·33286	129	126·0	...	3·0
95–	0·43289	18	17·3	...	0·7
100–	...	1	1·5	0·5	...
		14 480	14 492·1	115·8	103·7
				219·5	

5. It will be interesting to examine a particular case of the method just described, as it is often required by actuaries.

Defining Makeham's hypothesis as $\operatorname{colog} p_x = A + Bc^x$, we take a normal curve $(y_0 e^{-(x-h)^2/2\sigma^2})$ to represent the exposed and multiply by the values of $\operatorname{colog} p_x$. This means that we assume that the products can be represented by

$$\begin{aligned} y &= (A+Bc^x)y_0 e^{-(x-h)^2/2\sigma^2} \\ &= Ay_0 e^{-(x-h)^2/2\sigma^2} + By_0 e^{-(x^2-2hx+h^2-2\sigma^2 x \log_e c)/2\sigma^2} \\ &= Ay_0 e^{-(x-h)^2/2\sigma^2} + HBy_0 e^{-(x^2-2[h+\sigma^2 \log_e c]x+[h+\sigma^2 \log_e c]^2)/2\sigma^2} \end{aligned}$$

where $H = e^{\{h^2+2\sigma^2 h \log_e c+\sigma^4(\log_e c)^2-h^2\}/2\sigma^2} = e^{h \log_e c + (\sigma^2/2)(\log_e c)^2}$

$$\therefore \quad y = Ay_0 e^{-(x-h)^2/2\sigma^2} + HBy_0 e^{-(x-t)^2/2\sigma^2} \qquad \text{......(1)}$$

i.e. the sum of two normal curves both having the same standard deviation as the exposed curve and one having the same origin.

The difference between the two means gives $\sigma^2 \log_e c$, so

$$\log_{10} c = ((t-h)/\sigma^2)\log_{10} e.$$

The whole solution is made very simple by taking moments about the known origin (age h), for $\int_{-\infty}^{+\infty} xy\,dx$ and $\int_{-\infty}^{+\infty} x^2 y\,dx$ (the first two moments) give

$$(t-h)N_2\,^* \text{ and } N_1\sigma^2+N_2\{\sigma^2+(t-h)^2\}\,^\dagger$$

where $N_1 = Ay_0\sigma\sqrt{(2\pi)}$ and $N_2 = HBy_0\sigma\sqrt{(2\pi)}$.

Dividing the values just given by N_1+N_2 (the total frequency), we obtain, as the first moment about the known origin, $\dfrac{(t-h)N_2}{N_1+N_2}$, and, as the second,

$$\frac{N_1\sigma^2+N_2\sigma^2+N_2(t-h)^2}{N_1+N_2} = \sigma^2+(t-h)\mu_1'$$

or

$$t-h = \frac{\mu_2'-\sigma^2}{\mu_1'}$$

and

$$N_2 = \frac{\mu_1'(N_1+N_2)}{t-h}$$

where μ' is written for moments about h.

As stated above

$$\log_{10} c = \frac{t-h}{\sigma^2}\log_{10} e \qquad \text{......(2)}$$

and if $y_0 = \dfrac{10^k}{\sigma\sqrt{(2\pi)}}$ as is generally convenient, then

$$A = N_1/10^k$$

and

$$\begin{aligned} B &= N_2/(10^k \times H) \\ &= \frac{N_2}{10^k}\cdot\frac{1}{e^{h\log_e c+(\sigma^2/2)(\log_e c)^2}} \\ &= \frac{N_2}{10^k c^h e^{[(t-h)/2]\log_e c}} \qquad \text{(see equation (3))} \\ &= \frac{N_2}{10^k c^{(t+h)/2}} \end{aligned}$$

Care is necessary with regard to the value used for y_0, and consequently with regard to A and B. If Sheppard's tables in *Biometrika Tables* of ordinates (z) be multiplied by, say, 10^5 and used as the

* Remember that the normal curve is symmetrical, so that the odd moments about the mean of such a curve are zero.

† Can be seen at once as the sum of two integrals; $N_1\sigma^2$ gives the second moment of the first normal curve in (I), and $N_2\{\sigma^2+(t-h)^2\}$ gives the second moment of the second normal curve.

exposed to risk, the values of A and B resulting from the work will be $N_1/(10^5\sigma)$ and $N_2/(10^5 H\sigma)$. The reason is that his tables are in terms of standard deviation.

6. If we assume, as Hardy did when graduating the British Offices 1863–1893 experience, that $\log_{10} c$ is known, we only require to calculate one moment which gives us $(t-h)N_2/(N_1+N_2)$, and this, with the help of equation (2), enables us to complete the solution. If c were obtained for the aggregate table, we should use this result for the select tables.

7. A numerical example with the $O^{NM(5)}$ Table may be of interest. A normal curve with standard deviation 10 and origin $55\frac{1}{2}$ was taken, and the terms multiplied by $\operatorname{colog} p_x$. These were then grouped in fives, and the first two moments calculated about age $55\frac{1}{2}$. One little point should be borne in mind in connection with the grouping though the centre of the base on which the product ($q_x \times$exposed) stands is $x+\frac{1}{2}$, the result ($\operatorname{colog} p_x \times$exposed) is an ordinate at x; the centre point of five ages 20 to 24 is $22\frac{1}{2}$ when q_x is used and 22 when $\operatorname{colog} p_x$ is used.

The figures were

$$N_1+N_2 = 136387$$

$$\text{1st moment about } 55\tfrac{1}{2} \text{ in 5-years unit} = 1{\cdot}416184$$

$$\text{2nd moment about } 55\tfrac{1}{2} \text{ in 5-years unit} = 4{\cdot}1929354$$

Deducting Sheppard's adjustment of $\frac{1}{12}$ from the second moment * and multiplying the first moment and the adjusted second moment by 5 and 25 respectively to make the unit one year instead of 5 years, we have

$$\mu_1' = 7{\cdot}080920$$

$$\mu_2' = 164{\cdot}384085$$

then

$$\log(t-h) = 0{\cdot}9586889$$

$$t-h = 9{\cdot}092617$$

$$\log_{10} c = 0{\cdot}03948873$$

$$A = 0{\cdot}00301749$$

$$B = 0{\cdot}00004518782$$

$$\log_{10} B = \bar{5}{\cdot}6550214$$

* As we are dealing with the sum of five ordinates in each group and not with an area, we should not, strictly speaking, use Sheppard's adjustment, but should deduct 0·08. The difference is small and the constants have not been recalculated. The formulae would be $\mu_2 = \nu_2 - 0{\cdot}08$ and $\mu_4 = \nu_4 - 0{\cdot}48\nu_2 + 0{\cdot}02752$ where μ is adjusted and ν unadjusted.

q_x was then calculated from the graduated colog p_x obtained from the values of A, B and c, and the following table of expected deaths was worked out. The values of q_x are given in the table showing the frequency curve graduation:

Age group	Graduated q_x for central age of group	Expected deaths	Actual deaths	Deviation +	Deviation −
Under 25	...	13·0	9	4·0	...
25–29	0·00812	67·0	69	...	2·0
30–	0·00882	211·6	205	6·6	...
35–	0·00991	380·8	369	11·8	...
40–	0·01162	566·9	588	...	21·1
45–	0·01431	799·7	801	...	1·3
50–	0·01854	1 057·5	1 064	...	6·5
55–	0·02517	1 392·7	1 399	...	6·3
60–	0·03551	1 790·2	1 752	38·2	...
65–	0·05160	2 153·0	2 164	...	11·0
70–	0·07639	2 249·3	2 216	33·3	...
75–	0·11415	1 888·7	1 965	...	76·3
80–	0·17053	1 213·6	1 237	...	23·4
85–	0·23352	519·1	494	25·1	...
90–	0·36484	136·6	129	7·6	...
95–	...	20·6	19	1·6	...
...	...	14 460·3	14 480	128·2	147·9
				276·1	

This result is very like that given by the late Sir G. F. Hardy, but avoids having to obtain c by trial. Hardy's expected and actual deaths balance better than the above, but I do not think the rates have been understated systematically; the 75–79 group accounts for the disagreement. The total deviation is less than Hardy's.

8. If we know the range of a curve we need not even with Type I find as many as four moments, for the equations on page 57, giving the moments about the start of the curve, afford a simple solution. We have

$$\mu_1' = \frac{b(m_1+1)}{m_1+m_2+2} \quad \text{and} \quad \mu_2' = \frac{b^2(m_1+1)(m_1+2)}{(m_1+m_2+2)(m_1+m_2+3)}$$

and writing $$\gamma_1 = \frac{\mu_1'}{b} \quad \text{and} \quad \gamma_2 = \frac{\mu_2'}{\mu_1' b}$$

we have
$$m_1+1 = \frac{\gamma_1(\gamma_2-1)}{\gamma_1-\gamma_2}$$

and
$$m_2+1 = \frac{(\gamma_2-1)(1-\gamma_1)}{\gamma_1-\gamma_2}$$

where μ' is written for a moment about the start of the curve.

9. If, however, we can only fix by general considerations the start of the curve, the following solution depending on three moments is of use.

Writing
$$\lambda_2 = \frac{\mu_2'}{\mu_1'^2} \quad \text{and} \quad \lambda_3 = \frac{\mu_3'}{\mu_2'\mu_1'}$$

the values of the constants in the equation to the curve are given by
$$m_1+1 = \frac{2(\lambda_2-\lambda_3)}{2\lambda_3-\lambda_2-\lambda_2\lambda_3}$$
$$m_2+1 = \frac{2(\lambda_2-\lambda_3)(\lambda_3-1)(1-\lambda_2)}{(2\lambda_3-\lambda_2-\lambda_2\lambda_3)(1+\lambda_3-2\lambda_2)}$$
$$b = \mu_1'\frac{m_1+m_2+2}{m_1+1}$$

and
$$a_1/a_2 = m_1/m_2$$

10. We may return to Type I for an example of the method of § 8 where we will assume that the curve starts at age 17·5 and has a range of 15·5 units. Considering the line for age 22 in the table on page 53 we see that 4·175 and 14·634 give S_2 and S_3, excluding the first group, and the moments about age 17 are then found to be 4·175 and 25·093; transferring to 17·5, we have 4·075 and 24·268; adding the moments for the first group, $0{\cdot}034\times\frac{1}{5}$ and $0{\cdot}034\times(\frac{1}{5})^2$ respectively, $\mu_1' = 4{\cdot}0818$ and $\mu_2' = 24{\cdot}26936$.

Hence
$$m_1 = 0{\cdot}3498 \qquad a_1 = 1{\cdot}735$$
$$m_2 = 2{\cdot}7758 \qquad a_2 = 13{\cdot}765$$
$$y_0 = 154{\cdot}2$$

and the mode is $17{\cdot}5+1{\cdot}735\times 5 = 26{\cdot}175$.

From these values the graduated figures for the first four groups are 37, 140, 152, 143.

11. We can use this method with Example 1, page 5. With no moment adjustment we obtain
$$y = 0{\cdot}89082(x-1{\cdot}02897)^{-{\cdot}629685}(26{\cdot}52626-x)^{1{\cdot}624275}$$

This indicates that the first term (308) should be assumed to be at less than unit distance before the second term (200). Recalculating the first two moments by assuming the distance between the first two terms to be 0·75 and fixing the range at 25 and start at 1·1, I found

$$y = 0{\cdot}89055(x-1{\cdot}1)^{-{\cdot}61830}(26{\cdot}1-x)^{1{\cdot}63900}$$

The graduation by this curve is shown in the following table:

Duration	Withdrawals	Graduated by Type I curve
1	308	315
2	200	192
3	118	100
4	69	74
5	59	57
6	44	46
7	29	38
8	28	31
9	26	26
10	21	22
11	18	19
12	18	16
13	12	13
14	11	11
15	5	9
16	11	7
17	7	6
18	6	5
19	1	4
20	3	3
21	1	2
22	3	2
23	2	1
24	...	1
	1000	1000

12. The calculation of the graduated area of the first group may present a difficulty, as a quadrature formula cannot be applied, and the following method gives a good way of obtaining a correct value

$$\int_0^x y_0' x^{m_1}(b-x)^{m_2}dx$$

$$= \int_0^x y_0' x^{m_1}\left(b^{m_2}-m_2 b^{m_2-1}x+\frac{m_2(m_2-1)}{2}b^{m_2-2}x^2-\dots\right)dx$$

$$= y_0' x^{m_1+1} b^{m_2} \left(\frac{1}{m_1+1} - \frac{m_2 x}{b(m_1+2)} + \ldots \right)$$

which is a rapidly convergent series when x is small. y_0' must be calculated by the formula

$$\frac{N}{b^{m_1+m_2+1}} \cdot \frac{\Gamma(r)}{\Gamma(m_1+1)\Gamma(m_2+1)}$$

An expression for finding the area of the first group in Type III curves is

$$\int_0^x y_0' e^{-\gamma x} x^p dx = y_0' x^{p+1} \left(\frac{1}{p+1} - \frac{\gamma x}{p+2} + \ldots \right)$$

where $y_0' = N\gamma^{p+1}/\Gamma(p+1)$.

6. Curves represented by series expansions

1. In the previous chapter we dealt with Pearson's system of frequency curves, but other methods have been used to describe frequency distributions. We have already seen that Pearson's system of curves describes the facts that have been collected about a variety of subjects connected with chance. A system is useless if it does not give approximately the distributions that actually occur. The binomial series is justified from this point of view as a description of the number of times events happen, because we have found from experience that the numbers given by it are realized approximately by trial. When we consider the matter we are almost compelled to admit that the real justification of any theory of probability is that events happen in the way such a theory leads one to expect, and if we wish to compare the systems of frequency curves that have been suggested in recent years, it should be done not so much by examining the ways in which they have been derived as by seeing what classes of distribution they represent and by noticing carefully the cases of failure and the difficulties of application.

2. As we know from experience that the binomial series actually represents a simple type of probability, it is natural to start from it and treat it, or its limits, as a part of any system; it must, in fact, be a special case of any more general type that may be evolved.

We can proceed either by building up a curve on assumptions which it seems natural to adopt or by taking a more complex series than the binomial (e.g. the hypergeometrical) and in either case an expression might be reached having greater generality than the binomial. But it must be remembered that the ultimate justification of any evolved formula rests mainly on its breadth of application to statistics which may reasonably be described as chance distributions. Such application is an important test of the fundamental assumptions that were adopted in reaching the formula, for it must be admitted that the plausibility of the initial statements would be poor defence of a curve which broke down whenever it was put to a practical test.

The well-known 'normal curve of error', with which we dealt on page 70, was a first step towards finding a simple frequency curve, but though it works well as a description of the binomial $(p+q)^n$

when p is approximately equal to q or when n is large, it is unsatisfactory in other cases. In actuarial work these cases frequently arise. At the ages attained by the majority of lives assured in any assurance office the rate of mortality or probability of a person dying in a year is small and the frequency distribution giving the number of deaths happening in successive years out of 50 cases, say, when $q = 0{\cdot}02$ and $p = 0{\cdot}98$, would not be satisfactorily described by the normal curve of error. It is true in a sense that the 'normal curve' is a law of great numbers, but if it can only deal with cases resting on such a basis it cannot have a large sphere of action in practical statistics and it can hardly be expected to be of value when a series is more like the hypergeometrical than the binomial.

3. It is this failure of the 'normal curve' that has led to the work of Pearson, Thiele, Charlier, Edgeworth, Bruns, Kapteyn and others, and the curves suggested by these writers are of considerable interest to all students of statistical mathematics. In this chapter, and the next one, we shall indicate how far some of these curves fit the statistics that arise in practice; how far, in fact, they graduate the rough figures obtained from the collected facts, and where they break down.

Before proceeding, however, it will be necessary to discuss briefly the suggested types. We may also mention an old difficulty in practical work of this nature, namely, that statistics are seldom obtained from strictly homogeneous material. This fact must be taken as one of the typical elements in practice, and if a series can graduate in spite of a small amount of heterogeneity it is, from some points of view, all the more valuable in much of the work that comes to the hands of an actuary or statistician.

4. We may now turn to an expression which we will call **Type A**, obtained by expansion in terms of the normal frequency curve and its derivatives. Taking the mean at zero (or equivalently, using deviations from the mean), the frequency curve is

$$F(x) = \phi_0(x) - \frac{1}{3!}\frac{\mu_3}{\sigma^3}\phi_3(x) + \frac{1}{4!}\left(\frac{\mu_4}{\sigma^4}-3\right)\phi_4(x) - \frac{1}{5!}\left(\frac{\mu_5}{\sigma^5}-\frac{10\mu_3}{\sigma^3}\right)\phi_5(x)+\ldots$$

$$= \sum_{j=0}^{\infty} A_j\phi_j(x)$$

where $\phi_0(x) = \dfrac{N}{\sigma\sqrt{(2\pi)}}e^{-x^2/2\sigma^2}$ and $\phi_n(x) = \sigma^n\dfrac{d^n}{dx^n}\phi_0(x)$

So that, if $\sigma = 1$, i.e. if we measure in terms of the standard deviation,

$$\begin{aligned}\phi_3(x) &= (3x-x^3)\phi_0(x)\\ \phi_4(x) &= (x^4-6x^2+3)\phi_0(x)\\ \phi_5(x) &= (-x^5+10x^3-15x)\phi_0(x)\end{aligned}$$

Generally $\phi_m(x) = (-1)^m H_m(x)\phi_0(x)$, where $H_m(x)$ is a polynomial of degree m (the coefficient of x^m being 1) known as the mth *Hermite* polynomial.

In applying these expressions x is measured throughout from the mean in terms of the standard deviation. It may be mentioned that the coefficients in round brackets in the equation for Type A as set out above are the third, fourth and fifth semi-invariants.

In *Tables for Statisticians* (Pt II, Tables V–VII)

$$\tau_{n+1}(h) = \frac{(-1)^n}{\sqrt{(n+1)!}} \cdot \frac{d^n}{dh^n}\left(\frac{1}{\sqrt{2\pi}} e^{-h^2/2}\right)$$

and when using these tables we write F as

$$\frac{N}{\sigma}\{\tau_1(x)+0{\cdot}81649658\sqrt{\beta_1}\,.\,\tau_4(x)+0{\cdot}45643546(\beta_2-3)\tau_5(x)+\ldots\ldots\} \qquad \ldots\ldots(5)$$

This series has been discussed by many writers, expecially on the Continent,* and it may be regarded as the use of the 'normal curve' as a generating function. It has, naturally, a greater range of applicability than the 'normal curve', but it is not of service in the more extremely skew cases, and it has been suggested by C. V. L. Charlier that, in such circumstances, an expression **Type B** should be used. This is

$$F(x) = B_0\psi(x)+B_1\psi^{\mathrm{I}}_{(x)}+B_2\psi^{\mathrm{II}}_{(x)}+\ldots$$

where $$\psi(x) = e^{-m}\frac{\sin \pi x}{\pi}\left[\frac{1}{x}-\frac{m}{1!(x-1)}+\frac{m^2}{2!(x-2)}-\ldots\right]$$

and $\psi^{\mathrm{I}}_{(x)} = \psi(x)-\psi(x-1)$, i.e. $\Delta\psi(x-1)$ and values of $\psi(x)$ for $x<0$ are assumed to be zero. Similarly $\psi^{N}_{(x)} = \Delta\psi^{N-1}_{(x-1)}$.

In the limit when m is an integer $\psi(x)$ becomes $e^{-m}m^x/x!$. This expression is already well-known in the theory of probability as Poisson's series—the 'normal curve' is sometimes spoken of as a 'law of great numbers' and the Poisson series as a 'law of small numbers'.

* Gram, Thiele, Charlier, Bruns, etc. In a memoir entitled *Researches into the Theory of Probability* (Meddelanden Lunds Astronomiska Observatorium, 1906), C. V. L. Charlier gives several numerical examples and many useful notes. J. P. Gram, on page 94 of *Om Rækkendriklinger, bestemte ved mindste Kvadraters Methode* (Copenhagen, 1879), says that Oppermann had suggested the formula some time before.

Type B uses $e^{-m}m^x/x!$ as a generating function similarly to the way in which Type A uses the 'normal curve'.

5. The fitting of Type B presents certain special difficulties as alternative methods are available, but we may as a preface to them point out that if we calculate the moments of $e^{-m}m_x/x!$ using all integral values of x from $x = 0$ to $x = \infty$ we obtain

$$\mu_2 = m \qquad \mu_3 = m \qquad \mu_4 = 3m^2+m$$

or

$$\beta_1 = \beta_2-3 = 1/m$$

This, however, assumes a system of ordinates, unit distance apart, and we know that in practical statistical work these assumptions limit us unduly.

We can, however, write

$$F(xw+c) = B_0\psi(x)+B_1\psi^{\mathrm{I}}_{(x)}+B_2\psi^{\mathrm{II}}_{(x)}+\ldots$$

which implies that owing to w we have generalized the unit of grouping and owing to c the point from which x is reckoned is also generalized.

In this form Charlier suggests four methods of fitting and remarks that the series usually becomes more convergent if we arrange constants so that $B_1 = B_2 = B_3 = 0$.

(1) Assume $w = 1$ and $c = 0$, that is, revert to the original form and choose m so that B_1 vanishes, and since $B_0 = N$ we can reach

$$\begin{aligned}
2!B_2 &= N(\mu_2-b)\\
3!B_3 &= N(-\mu_3+3\mu_2-2b)\\
4!B_4 &= N(\mu_4-6\mu_3-6b\mu_2+11\mu_2+3b^2-6b)
\end{aligned}$$

where b is the distance from the origin to the mean. This method can be used when we can anticipate that m will not differ greatly from b.

(2) Assume $w = 1$ and calculate c as an unknown constant, choosing it and m so that B_1 and B_2 vanish.

$$\begin{aligned}
c &= b-\mu_2 \qquad & 3!B_3 &= N(\mu_2-\mu_3)\\
m &= \mu_2 \qquad & 4!B_4 &= N(\mu_4-3\mu_2^2-6\mu_3+5\mu_2)
\end{aligned}$$

(3) Find m, w and c so that $B_1 = B_2 = B_3 = 0$.

$$\begin{aligned}
w &= \mu_3/\mu_2 \qquad & B_0 &= N/w\\
m &= \mu_2^3/\mu_3^2 \qquad & B_4 &= \frac{N}{24w^5}\left(\mu_4-3\mu_2^2-\frac{\mu_3^2}{\mu_2}\right)\\
c &= b-\mu_2^2/\mu_3
\end{aligned}$$

This method usually gives w very small values and m very large values when μ_3 vanishes, so it is only applicable in markedly skew cases.

(4) Fix c arbitrarily and find m and w so that $B_1 = B_2 = 0$.

$$\begin{aligned} m &= (b-c)^2/\mu_2 \\ w &= \mu_2/(b-c) \\ B_0 &= N/w \\ w^3 3!B_3 &= B_0(w\mu_2-\mu_3) \\ w^4 4!B_4 &= B_0(\mu_4-3\mu_2^2+5w^2\mu_2-6w\mu_3) \end{aligned}$$

It seems unnecessary to give the work in detail leading up to the various sets of equations. Tables of $e^{-m}m^x/x!$ will be found in *Biometrika Tables.*

6. F. Y. Edgeworth * has used a series similar to Type A, namely:

$$e^{-\frac{k_1}{3!}\left(\frac{d}{dx}\right)^3+\frac{k_2}{4!}\left(\frac{d}{dx}\right)^4-\dots}\phi_0(x)$$

where k_1, k_2, etc. are the third, fourth, etc. semi-invariants. Expanding the exponential, we reach

$$\frac{N}{\sigma}\{\tau_1(x)+0{\cdot}81649658\sqrt{\beta_1}\tau_4(x)+0{\cdot}98601330\beta_1\tau_7(x)+\dots$$
$$+0{\cdot}45643546(\beta_2-3)\tau_5(x)+\dots+\text{etc.}\} \qquad \dots\dots(6)$$

Arithmetically the difference between this series and Type A is usually small. So far as terms involving moments up to and including 4th order are concerned, the difference between (5) and (6) is only the inclusion of the term $\frac{1}{72}\beta_1(x^6-15x^4+45x^2-15)\phi_0(x)$ in the latter. Later terms would also differ, but the expansion shown assumes that we shall not use more than four moments and that τ_9 etc. terms can be ignored. Sometimes the two series are distinguished by describing Type A as Type Aa, and Edgeworth as Type Ab.

7. It is possible to use other expressions, e.g. Type III † instead of a normal curve as a generating function. A necessary condition for a frequency function is that it must not produce negative frequencies and the reader who wishes to pursue this part of the subject may be referred to a lecture by Professor Steffensen giving

* Edgeworth contended that his equation was unique in its character and theoretical basis. *Trans. Camb. Phil. Soc.* 1905 (Law of Error), *J. Roy. Statist. Soc.* 1906 (Generalized Law of Error).

† S. H. Khamis, *Bull. Intern. Statist. Inst.* XXXVII (1960), 385–96; also M. L. Tiku, *Biometrika*, LII (1965), 415–27.

an interesting account from first principles.* For a general discussion of Edgeworth's and the A series and the theory underlying them the reader should study the papers to which reference has already been made and also Professor H. Cramer's paper 'On the composition of elementary errors' in *Skandinavisk Aktuarietidskrift*, 1928, page 13 etc. and page 141 etc.

It is not, however, pretended that the curves and series set out above exhaust the suggestions that have been made, but they may be taken to represent the methods that have received most general support, and the examples we shall give here do not go beyond them. In Chapter 7 we will describe yet further types of curves, obtained by the 'Method of Translation'. This consists of supposing that a normal curve applies to some simple function of x.

8. NUMERICAL EXAMPLES

Example 1

(Symmetrical curve not capable of satisfactory graduation by the normal curve of error.)

Observations	Pearson's Type II	Type A	Edgeworth	Normal curve
11	14	15	16	20
116	109	106	106	95
274	286	284	285	270
451	433	437	436	456
432	433	437	436	456
267	285	283	284	270
116	109	106	106	95
16	14	15	16	20

In this case all the curves except the normal give excellent graduations. We have not used Type B because Charlier apparently only adopts it when Type A is unsuccessful. He does not give a statistical criterion to show when A or B should be used and it is difficult to see how such a criterion can be evolved. The solution of his Type A does not lead to imaginary quantities when Type B should have been used, in the way that Pearson's Type I, for example, does when it is inapplicable. In reaching Type A and the Edgeworth graduation we have used the expressions (5) and (6) respectively.

* J. F. Steffensen, *Some recent researches in the Theory of Statistics and Actuarial Science* (Cambridge University Press, 1930, Third Lecture).

Example 2

(A distribution which is not markedly skew)

Observations	Pearson's Type III	Type A	Edgeworth
3	4	5	4
20	17	22	17
38	42	47	42
63	59	60	59
51	53	50	53
29	33	27	32
21	15	13	15
4	5	4	6
0	1·4	1	2
1	0·4	...	1

In each case three moments have been used. The observations and Edgeworth's graduation are taken from Edgeworth's paper, *The Generalised Law of Error*. Type A is the least successful.

Example 3

(A distinctly skew distribution)

Observations	Pearson's Type I	Type A	Type B	Edgeworth
...	...	−2	...	...
...	...	1	...	1
...	...	8	...	9
...	2	25	12	30
64·0	67	53	64	64
116·0	116	90	104	102
140·0	138	125	129	130
145·0	139	145	134	135
134·0	128	143	128	130
106·0	110	123	116	111
82·0	89	93	93	92
72·0	69	65	73	73
49·0	51	44	53	53
37·0	35	31	36	36
25·0	24	23	25	20
13·0	15	16	14	10
10·0	9	10	10	4
5·0	5	5	5	...
2·0	2	2	2	...
0·4	1	1	1	...

Pearson's figures come from his *Chances of Death* * and Edgeworth's from his *Generalised Law of Error*. Each of these graduations was obtained with four moments. Clearly Pearson's Type I is the best and Type B the next best graduation. We do not think Charlier would use Type A in such a case. In fitting his Type B there are, however, many difficulties owing to the fact that he gives us four approximate methods of application; this is an objection which may be surmounted in the future, but makes Type B awkward at present. The other points to be noticed in these graduations are the negative frequency in Type A and the 40 cases in Edgeworth's graduation which have no case corresponding to them in the data. Edgeworth, however, has remarked that he only aims at the main body of the curve and does not much concern himself with the tails, but one cannot help feeling that the main body must be understated if one tail possesses an excess of 40 out of 1000 cases and the other tail is in defect by only 20.

Example 4

(J-shaped curve)

Observations	Pearson's Type I	Type B
133	136·9	134·9
55	48·5	51·6
23	22·6	22·5
7	9·6	9·5
2	3·4	2·9
2	0·8	0·6

The Type B curve is given by Charlier in *Researches into the Theory of Probability*. The Type B curve gives a slightly better graduation, but the agreement is close in both cases. The example is not conclusive as to J-shaped curves, but shows that Type B can graduate them successfully. The particular example has only six groups, and with a curve of something like the right shape and three constants we are likely to reach close agreement. Edgeworth's curve is unsuitable. A graduation by Type A has been given elsewhere, but though it apparently graduates the figures the curve is not J-shaped.

* *Chances of Death*, I, 74 (London, 1897).

Example 5

(Series which is nearly symmetrical)

Observations	Pearson's Type IV	Type A	Edgeworth
10	6	4	3
13	16	14	10
41	49	46	34
115	135	126	110
326	321	306	298
675	653	637	662
1113	1108	1108	1164
1528	1535	1563	1603
1692	1712	1753	1747
1530	1522	1548	1510
1122	1074	1075	1024
611	604	589	571
255	274	256	263
86	102	92	104
26	32	29	37
8	8	7	12
2	2	2	2
1	1	1	1
1	...	...	...

These graduations give similar results and need no comment.

Example 6

(Distribution having two maxima)

Data	Pearson's Type II	Type A	Edgeworth
10	3	26	4
78	96	74	34
193	191	156	135
286	261	262	270
303	304	354	363
291	319	390	390
303	304	354	363
286	261	262	270
193	191	156	135
78	96	74	34
10	3	26	4

This is an imaginary example giving a double-humped distribution. It was formed from Type A by putting $A_3 = 0$ and $A_4 = 0{\cdot}09$; the series being

$$-4,\ -19,\ -53,\ -76,\ +103,\ +783,\ +1929,\ +2855,\ \text{etc.}$$

The negative frequencies were discarded and the data cut down and graduated. The interesting feature is that Type A (from which the 'data' were formed) gives a poor agreement. This is due to the negative frequencies and the integration for moments from $-\infty$ to $+\infty$. Negative frequencies are somewhat objectionable in themselves; they are still more objectionable when they influence curve fitting to the large extent shown in this example.

9. The few examples we have given will be of help in bringing out the comparison of the types of curves with which we have been dealing.

The Pearson-type curves will graduate satisfactorily all the examples we have taken, but cannot reproduce the double hump of our imaginary data (Example 6). They will graduate symmetrical, slightly skew and very skew distributions and also **J** and **U**-shaped distributions. They have been fitted in various circumstances and are satisfactory from the point of view of agreement. The arithmetic involved is, however, very heavy, but the curves are the most useful of those now considered.

Type A gives numerically the least work, but it does not graduate satisfactorily very skew or **J** and **U**-shaped distributions and it has therefore a smaller vogue. If, however, it is combined with Type B as Charlier suggests, **J**-shaped and skew distributions can be graduated. We have found some difficulty in applying Type B, for Charlier does not give much help in deciding which of his four methods of fitting should be followed in a particular case.

Edgeworth's curve can, roughly speaking, graduate the same distributions as Type A.

10. We may now refer to two difficulties in connection with Edgeworth's curve and with Type A which have already been mentioned. In Example 3 we found that 40 out of 1000 in Edgeworth's graduation have no observations corresponding to them and we remarked that it seemed a large excess; the reproduction of the exact number of observations is not only a practical necessity, but is assumed by the method of moments. If, therefore, a large number of cases falls outside the observations, we must either say that the total frequency is not reproduced or that the frequencies are misplaced; in either case the main body must be artificially reduced below the amount shown in the original data. In slightly skew distributions the frequencies are satisfactorily reproduced and many of the graduations of such material are excellent, but the method can hardly be considered satisfactorily as a general formula until some method of overcoming the difficulty mentioned above has been found.

The other difficulty is the large part that negative frequencies play in some of the less symmetrical graduations. If a negative frequency occurs, have the positive frequencies been overstated? The defence of such negatives is that further terms of the series would put things right, but it is hard to see the justification for basing much argument on constants derived from the higher moments which are liable to large variations and are unreliable. It is also unsatisfactory that a curve cannot reproduce itself even approximately, and the result of our Example 6 is disappointing; probably however it would be well to consider such cases as relating to heterogeneous material and therefore more suitable for representation by two or more superimposed curves. If Type A or Edgeworth's curve and their moments could be integrated from $-a$ to b instead of from $-\infty$ to ∞, the difficulties could be overcome to some extent; but, failing that, it would seem necessary to limit the range of applicability to the less abnormal distributions. An approximate method of fitting from $-a$ to ∞ has been given*, but the results are not quite so good as Pearson's Type III.

If the reader makes any extensive trial with series for the purpose of graduation, he will find occasionally that the coefficients of successive terms are such as to imply that the series may not be convergent. This is closely connected with the difficulty mentioned in the preceding paragraph.

The diagram † below shows the regions in the (β_1, β_2) plane in which the Gram-Charlier (5) and Edgeworth (6) curves (using up to 4th moments only) are unimodal and/or always having positive ordinates. It can be seen that the regions are not large, and fall mainly in the Pearson Type IV region.

* E. C. Rhodes, *J. Roy. Statist. Soc.* 1925, pp. 576 et seq.

† Taken from D. E. Barton and K. E. R. Dennis, *Biometrika*, LII (1965), 426.

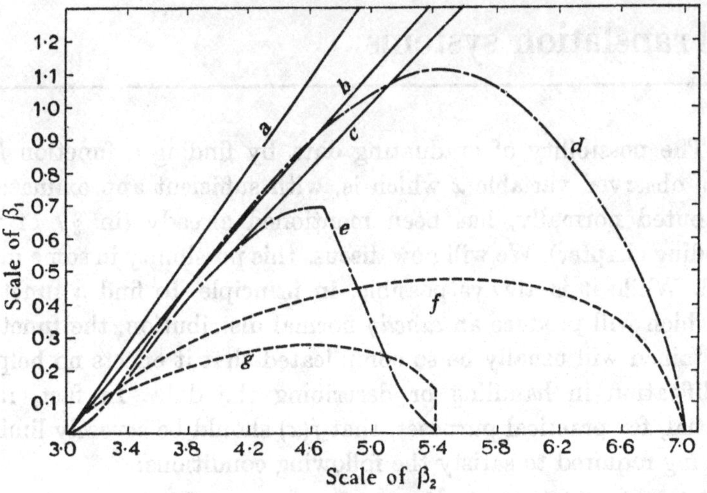

(*a*) Pearson Type III line; (*b*) Log normal line; (*c*) Pearson Type V line; (*d*) Gram-Charlier curves, boundary of positive definite region; (*e*) Gram-Charlier curves, boundary of unimodal region; (*f*) Edgeworth curves, boundary of positive definite region; (*g*) Edgeworth curves, boundary of unimodal region.

7. Translation systems

1. The possibility of graduating data by finding a function $f(x)$ of the observed variable x which is, with sufficient approximation, distributed normally, has been mentioned already (in §7 of the preceding chapter). We will now discuss this possibility in some more detail. While it is always possible, in principle, to find a function $f(x)$ which will produce an *exactly* normal distribution, the function so obtained will usually be so complicated that it effects no helpful simplification in handling or describing the data. In fact, it is essential, for practical purposes, that $f(x)$ should be severely limited by being required to satisfy the following conditions:

(*a*) it should be a simple function, expressible in terms of few arbitrary parameters.

(*b*) it should enable as wide as possible a variety of distributions to be described. Apart from this, it should evidently be a monotonic function of x.

If $f(x)$ is normally distributed, then it is possible to find values γ and δ such that

$$z = \gamma + \delta f(x)$$

is a standardized normal variable—i.e. z is normally distributed with expected value zero and standard derivation 1. (It can always be arranged that $\delta > 0$, and we will follow this convention.) The distribution of x, and in particular the values of β_1 and β_2 for this distribution, are defined by this equation. The values of β_1 and β_2 are thus determined by the values of γ and δ.

Two further parameters, ξ and λ, can be introduced by replacing x by $(x-\xi)/\lambda$, so that

$$z = \gamma + \delta f\left(\frac{x-\xi}{\lambda}\right)$$

Supposing that $f(\cdot)$ does not depend on any further arbitrary parameters we see that the distribution of x is determined by the values of the four parameters γ, δ, ξ, λ. Further, since the values of β_1 and β_2 are the same for $(x-\xi)/\lambda$ (or indeed, any non-constant linear function of x) as they are for x, they do not depend on ξ and λ—i.e. they are still determined by γ and δ.

By appropriate choice of the function $f(\cdot)$, therefore, we can construct systems of curves depending on four parameters (as do the curves of the Pearson system). The choice of $f(\cdot)$ is a matter of some importance. It is necessary to bear in mind the requirements (*a*) and (*b*) mentioned earlier in the present section. They are to some extent contradictory, in that it is found that if simple functions alone are considered, it is not possible to cover the whole of the (β_1, β_2) plane using a single function $f(\cdot)$. At least two (with a transitional form between them) are needed. We now proceed to describe such a pair of functions, forming a system covering the whole (β_1, β_2) plane.*

2. We first take $f(x) = \log_e\{x/(1-x)\}$ $(0 < x < 1)$. Replacing x by $(x-\xi)/\lambda$ we obtain a four-parameter system of curves defined by

$$z = \gamma + \delta \log_e\left[\frac{x-\xi}{\xi+\lambda-x}\right] \quad (\xi < x < \xi+\lambda)$$

with z a standardized normal variable. It is to be noted that the range of variation of x is *bounded*. To indicate this fact, the system of curves may be called the S_B *system*.

In order to represent distributions with unbounded range of variation, we need a function $f(x)$ which will be such that it is an increasing function of y for all values of x and tends to $\pm\infty$ as x tends to $\pm\infty$. Such a function is

$$f(x) = \sinh^{-1} x = \log_e\{x+\sqrt{x^2+1}\}$$

The four-parameter system of curves defined by

$$z = \gamma + \delta \sinh^{-1}\left(\frac{x-\xi}{\lambda}\right)$$

is called the S_U *system*, the suffix U indicating the unbounded range of variation of the variable.

If γ be kept fixed and δ increased without limit the corresponding curves, in both the S_U and S_B system, tend to the normal curve. If however δ increases without limit, with γ/δ remaining fixed, the limiting curve is of lognormal type. For lognormal curves,

$$z = \gamma + \delta \log_e(x-\xi)$$

has a standardized normal distribution. (Note that there are only *three* parameters ξ, x, δ here. If we write

$$z = \gamma' + \delta \log_e\left(\frac{x-\xi}{\lambda'}\right)$$

* See N. L. Johnson, *Biometrika*, XXXVI (1949), 149.

the λ' can be removed by writing $z = \gamma+\delta \log_e(x-\xi)$ with

$$\gamma = \gamma' - \delta \log_e \lambda'.$$

Lognormal curves form a transitional type between S_B and S_U curves.

We now proceed to more detailed discussion of these curves.

LOGNORMAL CURVES

$$y = N(\sqrt{2\pi})^{-1}\delta(x-\xi)^{-1}\exp[-\tfrac{1}{2}\{\gamma+\delta\log_e(x-\xi)\}^2] \quad (x > \xi)$$

Distribution of $z = \gamma+\delta\log_e(x-\xi)$ is described by standard normal curve ($y = (N/\sqrt{2\pi})\exp(\frac{1}{2}\gamma^2)$).

Origin ξ below start of curve.

The values to be calculated in order are

$$\delta = [\log_e\{(1+\tfrac{1}{2}\beta_1+\sqrt{\beta_1}(1+\tfrac{1}{4}\beta_1))^{\frac{1}{3}}+(1+\tfrac{1}{2}\beta_1-\sqrt{\beta_1}(1+\tfrac{1}{4}\beta_1))^{\frac{1}{3}}-1\}]^{-\frac{1}{2}}$$

($\omega = e^{\delta^{-2}}$ is the only real root of the equation $(\omega-1)(\omega+2)^2 = \beta_1$.)

$$\gamma = \tfrac{1}{2}\delta\log_e\left(\frac{\omega(\omega-1)}{\mu_2}\right)$$

$$\xi = \text{mean}-\exp(-\gamma\delta^{-1}+\tfrac{1}{2}\delta^{-2})$$

$$= \text{mean}-\frac{\sqrt{\mu_2}}{\omega-1}.$$

Median $= \xi+\exp(-\gamma\delta^{-1})$

Mode $= \xi+\omega^{-1}\exp(-\gamma\delta^{-1})$

Mean $= \xi+\omega^{\frac{1}{2}}\exp(-\gamma\delta^{-1})$

Note that

$$\frac{\text{Mean}-\text{Mode}}{\text{Mean}-\text{Median}} = \frac{\omega^{\frac{1}{2}}-\omega^{-1}}{\omega^{\frac{1}{2}}-1} = 1+\omega^{-\frac{1}{2}}+\omega^{-1}$$

$$= 3 \quad \text{if } \delta \text{ is large.}$$

Notes

The range is from ξ to infinity.

For this distribution

$$\beta_1 = (\omega-1)(\omega+2)^2; \qquad \beta_2 = \omega^4+2\omega^3+3\omega^2-3$$

The (β_1, β_2) point moves along a line as ω varies. This line is called the *lognormal line*. As ω decreases to 1 (i.e. as δ tends to infinity) the point approaches the position (0, 3) corresponding to the normal curve. As ω increases, both β_1 and β_2 increase. The curvature of the line appears quite slight.

The two parts of the (β_1, β_2) plane separated by the lognormal line, correspond to S_U (larger β_2's) and S_B (smaller β_2's) curves. All lognormal curves are unimodal. The mode is at $\xi+\omega^{-1}e^{-\gamma/\delta}$, the median is $\xi+e^{-\gamma/\delta}$. If the value of ξ is known (quite often it is reasonable to assume that it is zero) then only two moments are needed to fit the curve.

From the equations

$$\text{mean} = \omega^{\frac{1}{2}}e^{-\gamma/\delta}; \qquad \mu_2' = \omega^2 e^{-2\gamma/\delta}$$

we obtain

$$\omega = \mu_2'/(\text{mean})^2$$

and so

$$\delta = [\log\{\mu_2'/(\text{mean})^2\}]^{-\frac{1}{2}}$$

Also

$$\gamma/\delta = \log[\omega^{\frac{1}{2}}/(\text{mean})].$$

Proof

If $z=\gamma+\delta\log_e(x-\xi)$ then $x=\exp(z-\gamma)/\delta$. Since z has a unit normal distribution, the rth moment of x about zero is

$$\begin{aligned}\mu_r' &= \frac{1}{\sqrt{(2\pi)}}\int_{-\infty}^{\infty}\exp[-\tfrac{1}{2}z^2+r(z-\gamma)/\delta]\,dz\\ &= \frac{e^{-r\gamma/\delta}}{\sqrt{(2\pi)}}\int_{-\infty}^{\infty}\exp[-\tfrac{1}{2}(z-r/\delta)^2+\tfrac{1}{2}r^2/\delta^2]\,dz\\ &= \exp[-r\gamma\delta^{-1}+\tfrac{1}{2}r^2\delta^{-2}]\end{aligned}$$

where

$$\begin{aligned}\mu_1' &= \exp[-\gamma\delta^{-1}+\tfrac{1}{2}\delta^{-2}] = \omega^{\frac{1}{2}}e^{-\gamma/\delta}\\ \mu_2' &= \omega^2 e^{-2\gamma/\delta}\\ \mu_3' &= \omega^{9/2}e^{-3\gamma/\delta}\\ \mu_4' &= \omega^8 e^{-4\gamma/\delta}\end{aligned}$$

and

$$\begin{aligned}\mu_2 &= e^{-2\gamma/\delta}\omega(\omega-1)\\ \mu_3 &= e^{-3\gamma/\delta}\omega^{3/2}(\omega-1)^2(\omega+2)\\ \mu_4 &= e^{-4\gamma/\delta}\omega^2(\omega-1)^2(\omega^4+2\omega^3+3\omega^2-3)\\ \sqrt{\beta_1} &= (\omega-1)^{\frac{1}{2}}(\omega+2)\\ \beta_2 &= \omega^4+2\omega^3+3\omega^2-3\end{aligned}$$

The equation of the curve is

$$\begin{aligned}y &= \frac{N\delta}{\sqrt{(2\pi)}(x-\xi)}\exp[-\tfrac{1}{2}\{\gamma+\delta\log_e(x-\xi)\}^2]\\ &= \frac{N\delta}{\sqrt{(2\pi)}\,e^{\frac{1}{2}\gamma^2}(x-\xi)^{1+\gamma\delta}}\exp[-\tfrac{1}{2}\delta^2\{\log_e(x-\xi)\}^2]\end{aligned}$$

From this equation we find that $dy/dx = 0$ at $x = \xi$ and also at $x = \xi+\omega^{-1}e^{-\gamma/\delta}$. This latter is the only mode of the curve.

The median corresponds to $z = 0$.

S_U CURVES

$$y = \frac{N}{\lambda\sqrt{2\pi}} \cdot \frac{1}{\sqrt{[1+((x-\xi)/\lambda)^2]}} \exp[-\tfrac{1}{2}\{\gamma+\delta \sinh^{-1}[(x-\xi)/\lambda]\}^2]$$

Distribution of $z = \gamma+\delta \sinh^{-1}[(x-\xi)/\lambda]$

$$= \gamma+\delta \log_e[(x-\xi)/\lambda+\sqrt{(x-\xi)^2/\lambda^2+1}]$$

is described by the standard normal curve ($y = (N/\sqrt{2\pi})\, e^{-\frac{1}{2}z^2}$).

Values of γ and δ have to be determined from

$$\beta_1 = \frac{\omega(\omega-1)\{\omega(\omega+2)\sinh 3\Omega+3 \sinh \Omega\}^2}{2(\omega \cosh 2\Omega+1)^3}$$

$$\beta_2 = \frac{\omega^2(\omega^4+2\omega^3+3\omega^2-3)\cosh 4\Omega+4\omega^2(\omega+2)\cosh 2\Omega+3(2\omega+1)}{2(\omega \cosh 2\Omega+1)^2}$$

where $\Omega = \gamma/\delta$; $\omega = \exp(\delta^{-2})$.

Sign of γ is opposite to that of μ_3.

Tables are available to assist the solution. (See also the iterative method described in the Notes.)

Knowing γ and δ, ξ and λ are calculated from

Mean $= \xi-\lambda\omega^{\frac{1}{2}} \sinh \Omega$

$\mu_2 = \frac{1}{2}\lambda^2(\omega-1)(\omega \cosh 2\Omega+1)$

Notes

The range is unlimited. The sign of γ is opposite to that of μ_3. The curve is always unimodal.

The region of the (β_1, β_2) plane corresponding to Pearson Type IV curve falls entirely within the region corresponding to S_U. The S_U region also includes part of the Type VI region.

There are tables * which give values of γ and δ (to three decimal places) corresponding to specified values of β_1 and β_2. For greater accuracy or if the values of β_1 and β_2 are not among those in the table, the following iterative method may be used.

The method is based on the fact that lines $\delta = 0$ in the (β_1, β_2) plane are very nearly straight lines joining the points

$$(0, \tfrac{1}{2}(\omega^4+2\omega^2+3)) \quad (\delta = 0)$$

and $$((\omega-1)(\omega+2)^2, \omega^4+2\omega^3+3\omega^2-3) \quad (\delta = \infty)$$

* *Biometrika*, LII (1965), pp. 547–58.

Starting from an initial values $\delta = \delta_{(1)}$ with $\omega_{(1)} = \exp(\delta^{-2}_{(1)})$ we find the value $\gamma_{(1)}$ which will give the correct value for β_2 by solving (for $m_{(1)}$) the quadratic equation:

$$\beta_2 - 3 = \frac{(\omega_{(1)}-1)[A_2(\omega_{(1)})m^2_{(1)}+A_1(\omega_{(1)})m_{(1)}+A_0(\omega_{(1)})]}{2(2m_{(1)}+\omega_{(1)}+1)^2}$$

where

$$\begin{aligned} m_{(1)} &= \omega_{(1)}\sinh^2(\gamma_{(1)}/\delta_{(1)}) \\ A_2(\omega) &= 8(\omega^3+3\omega^2+6\omega+6) \\ A_1(\omega) &= 8(\omega^4+3\omega^3+6\omega^2+7\omega+3) \\ A_0(\omega) &= \omega^5+3\omega^4+6\omega^3+10\omega^2+9\omega+3 \end{aligned}$$

(Actually it is not necessary to calculate $\gamma_{(1)}$ explicitly at this stage. Only the value of $m_{(1)}$ need be obtained.)

Having obtained the value $m_{(1)}$, we next calculate the value of β_1 from the formula

$$\beta_1 = \frac{m(\omega-1)[4(\omega+2)m+3(\omega+1)^2]^2}{2(2m+\omega+1)^3}$$

(putting $m = m_{(1)}$, $\omega = \omega_{(1)}$).

Since this is a function of $\delta_{(1)}$ and the required value of β_2, it may be written $\beta_{(1)}(\delta_{(1)}, \beta_2)$.

We then obtain a new initial value for δ, $\delta_{(2)}$ say, by solving the equation

$$\frac{\beta_1}{\beta_1(\delta_{(1)}, \beta_2)} = \frac{\beta_2-\frac{1}{2}(\omega^4_{(2)}+2\omega^2_{(2)}+3)}{\beta_2-\frac{1}{2}(\omega^4_{(1)}+2\omega^2_{(1)}+3)}$$

The new value, $\delta_{(2)}$, is then used in place of $\delta_{(1)}$ in a repetition of the cycle. Cycles are repeated until $\beta_1(\delta_{(j)}, \beta_2)$ is sufficiently close to β_1. There is no need to check the value of β_2 as this is kept correct throughout the calculations.

Later stages of the solution may be speeded by determining $\delta_{(j)}$ from equations of form ($j = 3$)

$$\frac{\beta_1-\beta_1(\delta_{(1)}, \beta_2)}{\beta_1(\delta_{(2)}, \beta_2)-\beta_1(\delta_{(1)}, \beta_2)} = \frac{\frac{1}{2}(\omega^4_{(3)}+2\omega^2_{(3)}+3)-\frac{1}{2}(\omega^4_{(1)}+2\omega^2_{(1)}+3)}{\frac{1}{2}(\omega^4_{(2)}+2\omega^2_{(2)}+3)-\frac{1}{2}(\omega^4_{(1)}+2\omega^2_{(1)}+3)}$$

though this is not essential.

All S_U curves are unimodal.

The median is at $\xi-\lambda \sinh(\gamma/\delta)$.

Proof

Since $z = \gamma+\delta \sinh^{-1}[(x-\xi)/\lambda]$ it follows that

$$\frac{x-\xi}{\lambda} = \sinh\left(\frac{z-\gamma}{\delta}\right) = \tfrac{1}{2}[e^{(z-\gamma)/\delta}-e^{-(z-\gamma)/\delta}]$$

and so the rth moment of $(x-\xi)/\lambda$ is

$$(\sqrt{2\pi})^{-1}\, 2^{-r}\int_{-\infty}^{\infty} \{e^{(z-\gamma)/\delta}-e^{-(z-\gamma)/\delta}\}^r e^{-\frac{1}{2}z^2}\,dz$$

$$=(\sqrt{2\pi})^{-1}2^{-r}\sum_{j=0}^{r}\binom{r}{j}(-1)^j\int_{-\infty}^{\infty} \exp\{-\tfrac{1}{2}z^2+(r-2j)(z-\gamma)/\delta\}\,dz$$

If r is even

$$\mu_r' = 2^{-(r-1)}\left[\sum_{j=0}^{\frac{1}{2}r-1}(-1)^j\binom{r}{j}\omega^{\frac{1}{2}(r-2j)^2}\cosh[(r-2j)\Omega]+(-1)^{\frac{1}{2}r}\tfrac{1}{2}\binom{r}{\frac{1}{2}r}\right]$$

If r is odd

$$\mu_r' = 2^{-(r-1)}\sum_{j=0}^{\frac{1}{2}(r-1)}(-1)^{j+1}\binom{r}{j}\omega^{\frac{1}{2}(r-2j)^2}\sinh[(r-2j)\Omega]$$

Inserting the values $r = 1, 2, 3, 4$ and performing the appropriate calculations the formulae for mean, standard deviation, β_1 and β_2 we obtained.

The mode is at the value of x satisfying the equation

$$\left(\frac{x-\xi}{\lambda}\right)\left[1+\left(\frac{x-\xi}{\lambda}\right)^2\right]^{-\frac{1}{2}}+\delta\left[\gamma+\delta\sinh^{-1}\left(\frac{x-\xi}{\lambda}\right)\right] = 0$$

Since the left-hand side of this equation is an increasing function of x, varying from $-\infty$ to $+\infty$, the equation has a single root, so that there is only one mode. The corresponding value of x is $\lesseqgtr \xi$ according as $\gamma \gtreqless 0$.

The median corresponds to $z = 0$.

Example

As an example we will once again use the data from page 60 in Chapter 5, to which a Type IV curve was fitted. For these data,

$$\text{Mean} = 44{\cdot}577234$$
$$\sqrt{\mu_2} = 10{\cdot}639088$$
$$\beta_1 = 0{\cdot}0053656 \quad (\sqrt{\beta_1} < 0)$$
$$\beta_2 = 3{\cdot}172912$$

Since $\sqrt{\beta_1} < 0$, γ must be positive. The tables are not very adequate for β_1 and β_2 so near to the normal values (0 and 3). They suggest that $\delta_{(1)} = 5$ might be a suitable initial value, so that

$$\omega_{(1)} = \exp(\tfrac{1}{25}) = 1{\cdot}0408108$$

This leads to the equation:

$$0{\cdot}1792912 \times 2(2m_{(1)} + 2{\cdot}0408108)^2$$
$$= 0{\cdot}0408108(132{\cdot}9777186m_{(1)}^2 + 170{\cdot}7311940m_{(1)} + 34{\cdot}7070841)$$

for $m_{(1)}$, whence

$$m_{(1)} = 0{\cdot}005735$$

leading to $\beta_1(\delta_{(1)}, 3{\cdot}172912) = 0{\cdot}0021382$

and $\omega_{(2)}^2 = 1{\cdot}0811987$

i.e. $\omega_{(2)} = 1{\cdot}0398062$

Repetition of the cycle leads to

$$m_{(2)} = 0{\cdot}0149411$$
$$\beta_1(\delta_{(2)}, 3{\cdot}172912) = 0{\cdot}0053794$$
$$\omega_{(3)}^2 = 1{\cdot}0812076$$
$$\omega_{(3)} = 1{\cdot}0398113$$

A further repetition leads to

$$m_{(3)} = 0{\cdot}0148932$$
$$\beta_1(\delta_{(3)}, 3{\cdot}172912) = 0{\cdot}053631$$

which may be judged sufficiently close to the required value of 0·0053656.

(It is interesting to note that if $\omega_{(3)}$, had been determined from $\omega_{(1)}$ and $\omega_{(2)}$ using the second method of the Notes, its value would have been the same.)

Now (and only now) we obtain the implicit values of γ and δ corresponding to $m_{(3)}$ and $\delta_{(3)}$.

$$\sinh^2(\gamma/\delta) = m_{(3)}/\omega_{(3)} = 0{\cdot}0143230$$
$$\sinh(\gamma/\delta) = 0{\cdot}119679$$
$$\gamma/\delta = 0{\cdot}119395$$
$$\exp(\delta^{-2}) = 1{\cdot}0398113$$
$$\delta^{-2} = 0{\cdot}0390392$$
$$\delta = 5{\cdot}061155$$
$$\gamma = 0{\cdot}119395 \times 5{\cdot}061155 = 0{\cdot}604277$$

Hence, using the standard deviation of the observed values,

$$\lambda = \frac{10{\cdot}639088}{0{\cdot}202970} = 52{\cdot}4170$$

and using the mean of the observed values

$$\xi = 44{\cdot}577234 + 52{\cdot}4170 \times 0{\cdot}122038$$
$$= 50{\cdot}9741$$

The curve is fitted by regarding the distribution of

$$z = 0{\cdot}604277 + 5{\cdot}061155 \sinh^{-1}\left(\frac{x - 50{\cdot}9741}{52{\cdot}4170}\right)$$

as represented by the standard normal curve

$$y = \frac{N}{\sqrt{(2\pi)}} e^{-\frac{1}{2}z^2}$$

In the calculations it is convenient to use columns with headings as shown below:

x (Boundary values)	$y = 0{\cdot}0190778 \times (x - 50{\cdot}9741)$	$z = 0{\cdot}604277 + 5{\cdot}061155 \times \sinh^{-1} y$	$\phi(z)$	$\Delta\phi(z)$	Fitted frequency $N\Delta\phi(z)$
				0·0006426	6
7·5	−0·82939	−3·2192	0·0006428		
				0·0016035	15
12·5	−9·73400	−2·8388	0·0022463		
.	.	.	.	.	.
.	.	.	.	.	.
.	.	.	.	.	.

The complete set of figures in the final column is shown in the Table on page 60. The close similarity between the Type IV and S_U figures is notable.

S_B CURVES

$$y = \frac{N}{\lambda\sqrt{(2\pi)}}\left[\left(\frac{x-\xi}{\lambda}\right)\left(1 - \frac{x-\xi}{\lambda}\right)\right]^{-1} \exp\left[-\frac{1}{2}\left\{\gamma + \delta \log_e \frac{x-\xi}{\xi+\lambda-x}\right\}^2\right]$$
$$(\xi < x < \xi+\lambda)$$

Distribution of

$$z = \gamma + \delta \log_e \frac{x-\xi}{\xi+\lambda-x}$$

is described by a normal curve ($y = N(\sqrt{2\pi})^{-1} e^{-\frac{1}{2}z^2}$).

Sign of γ is the same as that of $\sqrt{\beta_1}$. β_1 and β_2 of the distribution of x are complicated functions of γ and δ; so are the mean and standard deviation.

Tables are not, as yet, available to aid in the solution. Recourse may be had to equating percentile points of observed and fitted curve (see Example).

Notes

The range of variation is limited in both directions, from ξ to $\xi+\lambda$. At both extremes ($x = \xi$ and $x = \xi+\lambda$), y tends to zero and there is infinitely high contact (all derivatives also tend to zero).

Modal values are roots of the following equation in $x' = (x-\xi)/\lambda$.

$$2x'-1 = \delta\left(\gamma+\delta \log_e \frac{x'}{1-x'}\right)$$

This may have one or three real roots between 0 and 1. If there are three roots there are two modes and an antimode (minimum). Such curves are comparable with the U-shaped Type I and Type II curves of the Pearson system. If there is only one root, the curve is unimodal. Bimodality occurs if

$$\delta < 1/\sqrt{2} \quad \text{and} \quad |\gamma| < \delta^{-1}\sqrt{(1-2\delta^2)}-2\delta \tanh^{-1}\sqrt{(1-2\delta^2)}.$$

The limiting values $\gamma = 0, \delta = 1/\sqrt{2}$ correspond to $\beta_1 = 0$, $\beta_2 = 1{\cdot}87$.

Example

Until suitable tables are available, it will not be practicable to fit S_B curves using moments, except when ξ and λ are known. (In this case it is merely necessary to fit a normal curve to the values of $\log[(x-\xi)/(\xi+\lambda-x)]$.) It is, however, relatively straightforward to fit S_B curves by equating four selected percentile points (three only, if ξ is known). It is desirable, from the point of view of accuracy, that the points selected should be about the lower and upper 7%, and lower and upper 25% points (quartiles). This is not always possible in practice—as, in fact, in the example we shall consider.

If the observed percentages up to x_1, x_2, x_3, x_4 are P_1, P_2, P_3, P_4 respectively, we have to solve the four equations

$$z_i = \gamma+\delta \log\left(\frac{x_i-\xi}{\xi+\lambda-x_i}\right) \quad (i = 1, 2, 3, 4)$$

where $$P_i = (\sqrt{2\pi})^{-1}\int_{-\infty}^{z_i} e^{-\frac{1}{2}u^2}\,du$$

for ξ, λ, δ and γ.

A method of solution by iteration, which is fairly easy to apply, is as follows. First select an initial pair of values for ξ and λ. We then have the four equations for γ and δ:

$$z_i = \gamma + \delta f_i \quad (i = 1, 2, 3, 4)$$

with $f_i = \log_e[(x_i - \xi)/(\xi + \lambda - x_i)]$ having known values.

By taking pairs of equations, explicit solutions for γ and δ can be obtained. In particular if the values of obtained by solving the first and second, and second and third, and third and fourth equations are all the same, then there will be a single pair of values of and satisfying all four equations.

Although it is troublesome to obtain exact solutions, it is not difficult to obtain approximations quite good enough for practical work. It is possible to work throughout with common logarithms (to base 10), multiplying the resulting value of δ (δ', say) by 0·4343 to make it appropriate to natural logarithms, if so desired.

It will be appreciated that a similar method can be used to fit S_U curves, though in the case the method of moments can also be used with little trouble.

As an example we take the data of Example 1 on page 5. (A Type I curve has been fitted to this on p. 108). We will interpret 'curtate duration *m*' to mean 'duration *m* to $(m+1)$', so that the successive groups represent durations of 1–2, 2–3, ..., 23–24 years.

Out of 1000 observations there are 308 less than 2; 508 less than 3; 754 less than 6; and 950 less than 14. The proportions are 0·308, 0·508, 0·754 and 0·950 and the corresponding values of z are (from Sheppard's tables) −0·5015, 0·0201, 0·6871 and 1·6449. The equations to be solved are

$$-0{\cdot}5015 = \gamma + \delta \log_e\left(\frac{2-\xi}{\xi+\lambda-2}\right)$$

$$0{\cdot}0201 = \gamma + \delta \log_e\left(\frac{3-\xi}{\xi+\lambda-3}\right)$$

$$0{\cdot}6871 = \gamma + \delta \log_e\left(\frac{6-\xi}{\xi+\lambda-6}\right)$$

$$1{\cdot}6449 = \gamma + \delta \log_e\left(\frac{14-\xi}{\xi+\lambda-14}\right)$$

As initial values of ξ and λ it seems reasonable (on the basis of the observed range of values) to take $\xi = 1$, $\lambda = 24$ (so that $\xi+\lambda = 25$). The calculations are summarized in the following table in which we use the symbol δ' (rather than δ) because logarithms to base 10 are used. First, a locus of pairs of values of ξ and λ for which the values of δ' based on the first and second, and on the third and fourth equations are equal is found. We seek the pair of values on this locus for which the value of δ' based on the second and third equations, equals the common value of the other two estimates. Some slight adjustments in the parameter values may be made to effect closer agreement.

ξ	λ	δ' based on equations 1st and 2nd	2nd and 3rd	3rd and 4th
1·0	24·0	1·6283	1·4449	1·4683
1·1	24·0	1·5175	1·4048	1·4598
1·156	24·0	1·4546	1·3820	1·4550
1·1	22·0	1·5093	1·3848	1·3794
1·2	22·0	1·3978	1·3453	1·3728
1·223	22·0	1·3718	1·3361	1·3712
1·224	22·0	1·3707	1·3357	1·3711
1·28	20·5	1·3002	1·2977	1·2984
1·28	20·6	1·3020	1·2978	1·3034
1·285	20·55	1·2961	1·2962	1·3006
1·287	20·43	1·2934	1·2941	1·2945
1·286	20·43	1·2946	1·2945	1·2946

(So we take (i) $\xi = 1{\cdot}286$; $\lambda = 20{\cdot}43$; $\delta' = 1{\cdot}2946$ and (from equations) $\gamma = 1{\cdot}3641$.)

It usually happens that quite good agreement is obtained over a considerable range of values of the parameters. This means that there is no pressing need for refinement in the calculations. In particular, when there is a long 'tail', the exact position of the end-point can be varied considerably with little effect on the fitted curve. In the present case, for example, one might feel that the upper limit of $20{\cdot}43+1{\cdot}286 = 21{\cdot}716$ is inappropriate since there are observations up to 24. Starting from the table of calculations shown above we obtain the values (ii) $\xi = 1{\cdot}2$; $\lambda = 22{\cdot}8$; $\delta' = 1{\cdot}39$; $\gamma = 1{\cdot}495$.

The sets of fitted frequencies for the two fitted S_B curves are shown in the table below. From these figures the close similarity of the two curves is immediately apparent.

Group	Observed frequency	Graduated by S_B (i)	(ii)
1–2	308	308	307
2–3	200	200	198
3–4	118	113	116
4–5	69	76	79
5–6	59	56	57
6–7	44	43	44
7–8	29	35	35
8–9	28	29	28
9–10	26	24	24
10–11	21	20	20
11–12	18	17	17
12–13	18	15	14
13–14	12	13	12
14–15	11	11	10
15–16	5	10	9
16–17	11	8	8
17–18	7	7	6
18–19	6	6	5
19–20	1	4	4
20–21	3	3	3
21–22	1	1	2
22–23	3	...	1
23–24	2	...	0
	1000	999	999

(i) $\xi = 1{\cdot}286$; $\lambda = 20{\cdot}43$, $\delta' = 1{\cdot}2946$; $\gamma = 1{\cdot}3641$. (ii) $\xi = 1{\cdot}2$; $\lambda = 22{\cdot}8$; $\delta' = 1{\cdot}39$; $\gamma = 1{\cdot}495$.

3. S_U and Type IV curves with the same values of $\sqrt{\beta_1}$ and β_2 have very similar shapes. The analysis connected with S_U curves is rather simpler than that associated with Type IV, rendering the former an attractive substitute for the latter. On the other hand, the position is reversed in regard to S_B and Type I curves. Calculations for the latter are much facilitated by the *Tables of the Incomplete Beta-Function*, while no tables are yet available for the former.

Interesting comparisons among the above systems are given in a paper by E. S. Pearson and J. W. Tukey (see Appendix 8).

8. Frequency surfaces

1. In order to describe the joint variation of two measurements, a three-dimensional representation is used. Denoting the two measurements by x_1, x_2 the equation

$$y = f(x_1, x_2) \qquad (\text{or } y = f_{x_1, x_2})$$

represents a surface. If $f(x_1, x_2)$ is such that the volume under the surface and vertically above a region R in the (x_1, x_2) plane is the frequency of values of (x_1, x_2) in R (for any choice of R) the equation represents a *frequency surface.*

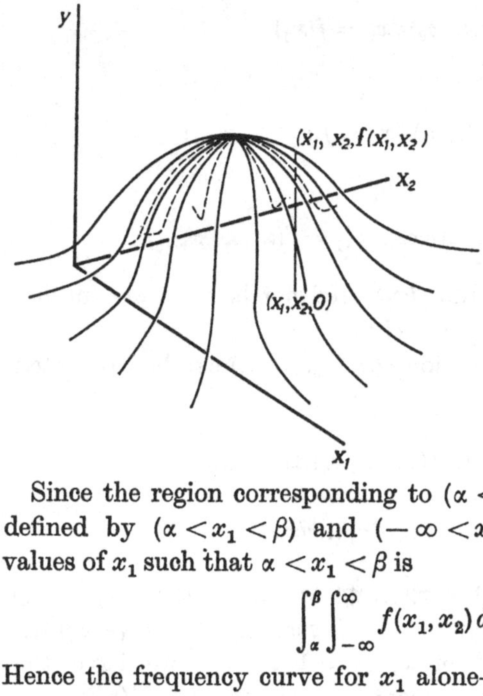

The Figure gives a diagrammatic representation of the situation.

Since the region corresponding to $(\alpha < x_1 < \beta)$ is an infinite slab defined by $(\alpha < x_1 < \beta)$ and $(-\infty < x_2 < \infty)$, the frequency of values of x_1 such that $\alpha < x_1 < \beta$ is

$$\int_\alpha^\beta \int_{-\infty}^{\infty} f(x_1, x_2)\, dx_2 dx_1$$

Hence the frequency curve for x_1 alone—often called the *marginal* frequency curve—is

$$\int_{-\infty}^{\infty} f(x_1, x_2)\, dx_2$$

Similarly the frequency curve for x_2 alone is

$$\int_{-\infty}^{\infty} f(x_1, x_2)\, dx_1$$

These ideas can be extended in a straightforward fashion to cases where are more than two measurements—x_1, $x_2, \ldots x_k$ say. In this case the frequency 'surfaces' $y = f(x_1, x_2, \ldots x_k)$ cannot be represented diagrammatically as in the Figure but we still have:

Frequency with which $(x_1, x_2, \ldots x_k)$ falls in the region R is equal to

$$\int \cdots \int_R f(x_1, \ldots x_k)\, dx_1 \ldots dx_k$$

(for any R).

(The notation $(x_1, x_2, \ldots x_k)$ indicates a point in k-dimensional space. '$(x_1, \ldots, x_k)$ in R' means that the values $x_1, \ldots, x_k$ satisfy certain conditions which define all the points in R.)

Also, it is possible to extend the relation

$$\int_{-\infty}^{\infty} f(x_1, x_2)\, dx_2 = f(x_1)$$

to

$$\int_{-\infty}^{\infty} f(x_1, x_2, \ldots, x_k)\, dx_k = f(x_1, \ldots, x_{k-1})$$

and, further

$$\int_{-\infty}^{\infty}\int_{-\infty}^{\infty} f(x_1, x_2, \ldots, x_k)\, dx_k\, dx_{k-1} = f(x_1, \ldots, x_{k-2})$$

and so on. (Note that $f(x_1)$ and $f(x_2)$ need not be the *same* mathematical functions of x_1, x_2 respectively.)

The mean value of any function $g(x_1, x_2, \ldots, x_n)$ can be calculated from the formula

$$\frac{\int \cdots \int g(x_1, \ldots, x_n) f(x_1, \ldots, x_n)\, dx_1 \ldots dx_n}{\int \cdots \int f(x_1, \ldots, x_n)\, dx_1 \ldots dx_n}$$

In particular if $g(x_1, \ldots, x_n) = x_1^{a_1} \ldots x_n^{a_n}$ we have the '$(a_1, \ldots, a_n)$ product moment about zero' and if $g(x_1, \ldots, x_n) = (x_1 - \xi_1)^{a_1} \ldots (x_n - \xi_n)^{a_n}$ where ξ_j = mean of x_j $(j = 1, 2, \ldots, n)$ we have the '$(a_1, \ldots, a_n)$ central product-moment' or simply the '$(a_1, \ldots, a_n)$ product moment'; usually denoted $p_{a_1, a_2, \ldots, a_n}$. If these quantities are divided by the product

$$(\text{s.d. of } x_1)^{a_1} \ldots (\text{s.d. of } x_n)^{a_n}$$

(where s.d. = standard deviation) the resulting scale-free quantities are called *reduced* product moments, usually denoted $q_{a_1, a_2, \ldots, a_n}$.

The (1, 1) (central) product moment of x_i and x_j is called their covariance (often written $\text{cov}(x_i, x_j)$). The corresponding reduced product moment is the *correlation coefficient* between x_i and x_j. Thus

$$\rho_{ij} = \text{correlation between } x_i \text{ and } x_j = \frac{\text{cov}(x_i, x_j)}{\sqrt{\text{var}(x_i)\,\text{var}(x_j)}}$$

(Note that, for example $\rho_{12} = p_{11}/\sqrt{(p_{20}p_{02})}$.)

The correlation is usually denoted by the symbol ρ, possibly with subscripts—e.g. ρ_{ij}—to indicate the variables concerned.

If x_2 be fixed, then $f(x_1, x_2)$ shows how x_1 varies for this value of x_2. Normalizing, by dividing by $\int f(x_1, x_2)\,dx_1 = f(x_2)$, we obtain the *conditional* frequency function of x_1, given x_2. The mean of this function is the *regression* of x_1 on x_2. Extension to the case of more than two variables follows a similar pattern.

2. Extension of the Pearson system of frequency curves to two dimensions has been studied by van Uven.* The generalization is affected by replacing the differential equation (1)′ on page 37 by the pair of partial differential equations.

$$\frac{\partial y}{\partial x_j} = \frac{L_j(x_1, x_2)}{Q_j(x_1, x_2)} \quad (j = 1, 2)$$

where L_j and Q_j are linear and quadratic functions, respectively, of the variables x_1 and x_2. It is clear that if one of the variables, x_1 say, is fixed, then $2y/2x_2$ satisfies an equation of form (1) and so the conditional distribution of x_2, given x_1 (and similarly of x_1, given x_2) is described by Pearson frequency curve.

The more important frequency surfaces are summarized in the table on page 139: the Type designations in the first column are those assigned by van Uven.

By equating two different expressions for $\partial^2 \log y/\partial x_1\,\partial x_2$, it can be seen that the functions L and Q cannot be chosen completely arbitrarily, but that they must satisfy the relationship

$$Q_2\frac{\partial L_1}{\partial x_2} - Q_1\frac{\partial L_2}{\partial x_1} = \frac{Q_2}{Q_1}L_1\frac{\partial Q_1}{\partial x_2} - \frac{Q_1}{Q_2}L_2\frac{\partial Q_2}{\partial x_1}$$

We further note from (6) that

$$\int Q_1\frac{\partial y}{\partial x_1}\,dx_1 = \iint Py\,dx_1$$

* M. J. van Uven, *Proc. Kon. Akad. Wetens., Amsterdam*, L (1947), 1063, 1252; LI (1948), 41, 191.

Integrating the left-hand side by parts we have

$$\Big[Q_1 y\Big]_{-\infty}^{\infty} - \int \frac{\partial Q_1}{\partial x_1} y\, dx_1 = \int Py\, dx_1$$

or

$$\int \left(P + \frac{\partial Q_1}{\partial x_1}\right) y\, dx_1 = \Big[Q_1 y\Big]_{-\infty}^{\infty}$$

Now $P + (\partial Q_1/\partial x_1)$ is a linear function of x_1 and x_2, and so the left-hand side is a linear function of x_2 and the mean of x_1, given x_2. The right-hand side is a function of x_2. If it is a linear function of x_2 (in particular, as in many cases, if it is zero) then it follows that the mean of x_1, given x_2 is a linear function of x_2; that is, the *regression* of x_1 on x_2 is *linear*.

Type	Equation $y =$	Conditions	Marginal types x_1	x_2
I	$f(x_1)f(x_2)$	(Independent variables with frequencies $f(x_1)$, $f(x_2)$)		
II$a\alpha$	$\dfrac{\Gamma(m_1+m_2+m_3)}{\Gamma(m_1)\Gamma(m_2)\Gamma(m_3)} x_1^{m_1-1} x_2^{m_2-1}(1-x_1-x_2)^{m_3-1}$	$m_1, m_2, m_3 > 0$; $x_1, x_2 > 0$; $x_1 + x_2 \leqslant 1$	I or II	I or II
II$a\beta$	$\dfrac{\Gamma(-m_3+1) x_1^{m_1-1} x_2^{m_2-1}(1+x_1+x_2)^{m_3-1}}{\Gamma(m_1)\Gamma(m_2)\Gamma(-m_1-m_2-m_3+1)}$	$m_1, m_2 > 0$; $m_1+m_2+m_3 < 0$; $x_1, x_2 > 0$	VI	VI
II$a\gamma$	$\dfrac{\Gamma(-m_2+1) x_1^{m_1-1} x_2^{m_2-1}(-1-x_1+x_2)^{m_3-1}}{\Gamma(m_1)\Gamma(m_3)\Gamma(-m_1-m_2-m_3+1)}$	$m_1, m_3 > 0$; $m_1+m_2+m_3 < 0$; $x_2 - 1 > x_1 > 0$	VI	VI
II$a\delta$	$\dfrac{\Gamma(-m_1+1) x_1^{m_1-1} x_2^{m_2-1}(-1+x_1-x_2)^{m_3-1}}{\Gamma(m_2)\Gamma(m_3)\Gamma(-m_1-m_2-m_3+1)}$	$m_2, m_3 > 0$; $m_1+m_2+m_3 < 0$; $x_1 - 1 > x_2 > 0$	VI	VI
IIb	$\dfrac{x_1^{m_1-1} x_2^{m_2-1} \exp[-(x_1+1)/x_2]}{\Gamma(m_1)\Gamma(-m_1-m_2)}$	$m_1 > 0$; $m_1+m_2 < 0$; $x_1, x_2 > 0$	VI	V
III$a\alpha$	$\dfrac{-m\sqrt{(1-\rho^2)}}{\pi k^m}(k + x_1^2 - 2\rho x_1 x_2 + x_2^2)^{m-1}$	$m < 0$; $\lvert\rho\rvert < 1$; $k > 0$	VII	VII
III$a\beta$	$\dfrac{m\sqrt{(1-\rho^2)}}{\pi k^m}(k - x_1^2 + 2\rho x_1 x_2 - x_2^2)^{m-1}$	$m > 0$; $\lvert\rho\rvert < 1$; $k > 0$; $x_1^2 - 2\rho x_1 x_2 + x_2^2 < k$	II	II
IVa	$\dfrac{x_1^{m_1-1}(x_2-x_1)^{m_2-1} e^{-x_2}}{\Gamma(m_1)\Gamma(m_2)}$	$m_1, m_2 > 0$; $0 < x_1 < x_2$	III	III
VI	$\dfrac{1}{2\pi\sqrt{(1-\rho^2)}} \exp\left[-\dfrac{1}{2(1-\rho^2)}(x_1^2 - 2\rho x_1 x_2 + x_2^2)\right]$	$\rho\rvert < 1$	Normal	Normal

In all the cases shown in the table, the regression of x_1 on x_2, and of x_2 on x_1, is, in fact, linear.

The marginal distributions of these frequency surfaces are represented by Pearson Type frequency curves, and it is possible to fit the parameters of these curves by moments in the way described in Chapter 5. The remaining parameters can be fitted by using the product moments of the observed sets of pairs of the variables.

Type II$a\alpha$, in which the marginal distributions are Pearson Type I (or II),* is also known as the *Dirichlet* family of distributions. For n variables, the Dirichlet family have frequency 'surfaces' with equations of form.

$$y = \frac{N\Gamma(m)}{\prod_{j=0}^{n} \Gamma(m_j)} x_1^{m_1-1} x_2^{m_2-1} \dots x_n^{m_n-1}(1-x_1-\dots-x_n)^{m_0-1}$$

where $m = m_0+m_1+\dots+m_n$. The formula holds for $0 < x_j$ $(j = 1, 2, \dots, n)$; $x_1+x_2+\dots x_n \leqslant 1$. Extra parameters are introduced by considering the distribution of $y_1, y_2, \dots y_n$ with

$$y_j = \sigma_j+\lambda_j x_j \, (\lambda_j > 0).$$

Type III$a\alpha$ may be regarded as a generalization of Pearson Type VII distributions. It is sometimes referred to a a 'bivariate t' distribution. For n variables there is the frequency surface:

$$y = [\Gamma(m)/\Gamma(m-\tfrac{1}{2}n)]\pi^{-\frac{1}{2}n}|A|^{\frac{1}{2}}\left[1+\sum_{i=1}^{n}\sum_{j=1}^{n} \alpha_{ij}(x_i-\xi_i)(x_j-\xi_j)\right]^{-m}$$

where $|A|$ denotes the value of the $n\times n$ determinant with elements α_{ij}. It is necessary that m exceed $\frac{1}{2}n$.

Type VI is a natural generalization of the normal frequency curve and is called a bivariate normal frequency surface. For n variables there is the *multinormal* frequency surface:

$$y = y_0 \exp\left[-\tfrac{1}{2}\sum_{i=1}^{n}\sum_{j=1}^{n} \alpha_{ij}(x_i-\xi_i)(x_j-\xi_j)\right]$$

where $y_0 = (2\pi)^{-\frac{1}{2}n}|A|^{\frac{1}{2}}$, the symbol $|A|$ again denoting the value of the $n\times n$ determinant with elements α_{ij}. It can always be arranged that $\alpha_{ij} = \alpha_{ji}$, so the determinant is symmetrical. Then the variance of x_i is $A^{ii}/|A|$, and the covariance of x_i and x_j is $A^{ij}/|A|$, where A^{ij} is the cofactor of α_{ij} in $|A|$. The mean of x_i is ξ_i.

We now give a more extended discussion of the bivariate normal surface and its applications.

BIVARIATE NORMAL SURFACE

3. The bivariate normal frequency surface is

$$y = \frac{N}{2\pi\sigma_1\sigma_2\sqrt{(1-\rho^2)}} \exp\left[-\frac{1}{2(1-\rho^2)}\left\{\left(\frac{x_1-\xi_1}{\sigma_1}\right)^2 - 2\rho\left(\frac{x_1-\xi_1}{\sigma_1}\right)\left(\frac{x_2-\xi_2}{\sigma_2}\right)+\left(\frac{x_2-\xi_2}{\sigma_2}\right)^2\right\}\right]$$

* But note that the two marginal curves cannot *both* be Type II. We have $f(x_1)$ proportional to $x_1^{m_1-1}(1-x_1)^{m_2+m_3-1}$, and $f(x_2)$ to $x_2^{m_2-1}(1-x_2)^{m_1+m_3-1}$ and we cannot have simultaneously $m_1 = m_2+m_3$ and $m_2 = m_1+m_3$.

ξ_1, σ_1 are the mean and standard deviation of x_1; ξ_2, σ_2 are the mean and standard deviation of x_2; ρ is the coefficient of correlation between x_1 and x_2. Putting $\xi_1 = \xi_2 = 0$; $\sigma_1 = \sigma_2 = 1$ we obtain the unit (standardized) bivariate normal surface

$$y = \frac{N}{2\pi\sqrt{(1-\rho^2)}} \exp\left[-\frac{1}{2(1-\rho^2)}(x_1^2 - 2\rho x_1 x_2 + x_2^2)\right]$$

It is convenient to use $\phi(x_1, x_2)$ to denote this function of x_1 and x_2, just as we can use

$$\phi(x) = \frac{N}{\sqrt{(2\pi)}} \exp(-\tfrac{1}{2}x^2)$$

to denote the standardized (or 'unit') normal frequency function. The equation of the general bivariate normal frequency surface can be written

$$y = (\sigma_1\sigma_2)^{-1}\phi\left(\frac{x_1-\xi_1}{\sigma_1}, \frac{x_2-\xi_2}{\sigma_2}\right)$$

If x_1 and x_2 have this frequency surface then the frequency surface for the standardized variables $x_1' = (x_1-\xi_1)/\sigma_1$; $x_2' = (x_2-\xi_2)/\sigma_2$ is the unit bivariate normal surface

$$y = \phi(x_1', x_2')$$

From Section 2 of Appendix 4 it can be seen that the marginal distribution of x_1' is described by the normal curve

$$y = \phi(x_1') = N(2\pi)^{-\frac{1}{2}} \exp(-\tfrac{1}{2}x_1'^2)$$

and that of x_2' is of the same form. It follows that for x_1:

$$y = \frac{N}{\sigma_1\sqrt{(2\pi)}} \exp\left[-\tfrac{1}{2}\left(\frac{x_1-\xi_1}{\sigma_1}\right)^2\right]$$

and for x_2:

$$y = \frac{N}{\sigma_2\sqrt{(2\pi)}} \exp\left[-\tfrac{1}{2}\left(\frac{x_2-\xi_2}{\sigma_2}\right)^2\right]$$

The distribution of x_2, given x_1, is described by the normal frequency curve with mean $\xi_2 + (\rho\sigma_2/\sigma_1)(x_1-\xi_1)$ (this is the regression function of x_2 on x_1) and standard deviation $\sigma_2\sqrt{(1-\rho^2)}$.

If it is desired to fit a bivariate normal surface to an observed distribution arranged in a two-way table, the parameters ξ_1, ξ_2, σ_1, σ_2 and can be estimated from the observed means $(\bar{x}_1, \bar{x}_2)$, standard deviations (s_1, s_2) and product-moment correlation (r). It is then necessary to calculate

$$(s_1 s_2)^{-1} \int_{a_2}^{a_2'} \int_{a_1}^{a_1'} \phi\left(\frac{x_1-\bar{x}_1}{s_1}, \frac{x_2-\bar{x}_2}{s_2}\right) dx_1\, dx_2$$

for values of a_1, a_1', a_2, a_2' corresponding to the group boundaries for x_1, x_2 respectively. This can be reduced to an integral of the standardized frequency function $\phi(x_1, x_2)$ alone. It is

$$\int_{h_2}^{h_2'}\int_{h_1}^{h_1'} \phi(x_1, x_2)\,dx_1\,dx_2$$

where $h_j = (a_j - \bar{x}_j)/s_j$; $h_j' = (a_j' - \bar{x}_j)/s_j$ $(j = 1, 2)$.

Using tables of

$$L(h, k; p) = \int_h^\infty \int_k^\infty \phi(x_1, x_2)\,dx_1\,dx_2$$

this can be evaluated as

$$L(h_1', h_2'; \rho) - L(h_1, h_2'; \rho) - L(h_1', h_2; \rho) + L(h_1, h_2; \rho)$$

Tables of $L(h, k; \rho)$ are available in *Tables for Statisticians* (Part II), and also in *Applied Math. Series No.* 50 (National Bureau of Standards, U.S. Govt. Printing Office).

Alternatively*, tables of the function

$$V(h, ah) = \frac{1}{2\pi}\int_0^h \int_0^{ax} e^{-\frac{1}{2}(x^2+y^2)}\,dx\,dy$$

may be used. Putting $a = \rho(1-\rho^2)^{-\frac{1}{2}}$ we have

$$V(h, ah) = \tfrac{1}{4} + L(h, 0; \rho) - L(0, 0; \rho) - \tfrac{1}{2} \cdot \frac{1}{\sqrt{(2\pi)}}\int_h^\infty e^{-\frac{1}{2}u^2}\,du$$

Values of $L(h, k; \rho)$ can be obtained by using the formulae

(*a*) $$L(h, k; \rho) = L(h, 0; \rho_{h,k}') + L(k, 0; \rho_{k,h}') - \delta$$

where $\rho_{h,k}' = (\text{sign of } h)(\rho h - k)(h^2 - 2\rho hk + k^2)^{-\frac{1}{2}}$

$$\delta^2 = \begin{cases} 0 & \text{if } hk > 0, \text{ or } hk = 0 \text{ and } h+k \geqslant 0 \\ -\frac{1}{2} & \text{otherwise} \end{cases}$$

(*b*) $L(0, 0; \rho) = 1 + \dfrac{\text{arc sin}\,\rho}{2}$ (Sheppard's formula).

Tables of $V(h, ah)$ are given in the reference in the footnote. More extensive tables (of $(2\pi)^{-1}$ arc tan $a - V(h, ah)$) are given by D. B. Owen in the reference given in Appendix 8.

4. The normal bivariate frequency surface can be used to estimate the correlation coefficient from the observed frequencies in a two-way table. It is supposed that each of the observed dichotomized variables are in reality such that their distribution can be represented by a unit normal frequency curve. (Since the values of the variable can be measured on any arbitrary linear scale without affecting the

* C. Nicholson, *Biometrika*, XXXIII (1943), 59–72.

value of the correlation coefficient, we lose nothing by requiring the mean to be zero, and the standard deviation to be 1, while gaining some algebraic convenience.) Then the frequencies a, b, c and d in the table at the foot of the page correspond to division of the surface

$$y = \frac{N}{2\pi\sqrt{(1-\rho^2)}} \cdot \exp\left[-\frac{1}{2(1-\rho^2)}(x_1^2 - 2\rho x_1 x_2 + x_2^2)\right]$$

into four parts by planes at right angles to the axes of x_1 and x_2 at distances h_1 and h_2 from the origin, as suggested by the upper figure on this page.

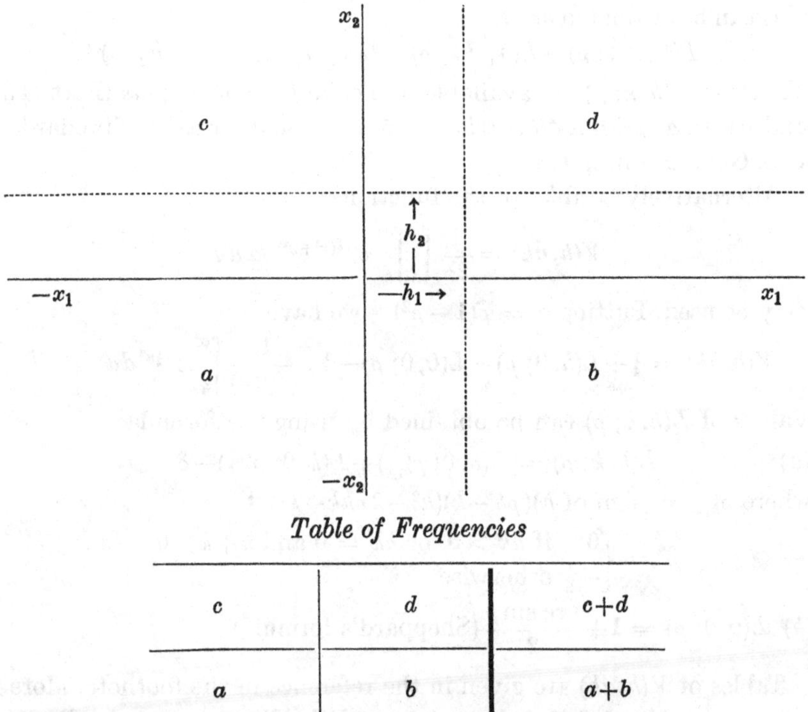

Table of Frequencies

c	d	$c+d$
a	b	$a+b$
$a+c$	$b+d$	N

Then

$$d = \frac{N}{2\pi\sqrt{(1-\rho^2)}} \int_{h}^{\infty}\int_{h_2}^{\infty} \exp\left[-\frac{1}{2(1-\rho^2)}(x_1^2 + x_2^2 - 2\rho x_1 x_2)\right] dx_1\, dx_2$$

Further

$$b+d = \frac{N}{\sqrt{(2\pi)}} \int_{h_1}^{\infty} e^{-\frac{1}{2}x^2}\, dx$$

and
$$c+d = \frac{N}{\sqrt{(2\pi)}}\int_{h_2}^{\infty} e^{-\frac{1}{2}x^2}\,dx$$
and, remembering that N (the total frequency) $= a+b+c+d$, we have
$$N-2(b+d) = N-N\sqrt{2/\pi}\int_{h_1}^{\infty} e^{-\frac{1}{2}x^2}\,dx$$
$$\therefore \qquad \frac{(a+c)-(b+d)}{N} = \sqrt{\frac{2}{\pi}}\int_0^{h_1} e^{-\frac{1}{2}x^2}\,dx$$
and, similarly
$$\frac{(a+b)-(c+d)}{N} = \sqrt{\frac{2}{\pi}}\int_0^{h_1} e^{-\frac{1}{2}x^2}\,dx$$
As a, b, c and d are known, h_1 and h_2 can be found from Sheppard's Tables, and the problem becomes

'To find a value for r from the equation
$$\frac{N}{2\pi\sqrt{(1-r^2)}}\int_{h_1}^{\infty}\int_{h_2}^{\infty}\exp\left[-\frac{1}{2(1-r^2)}(x_1^2-2rx_1x_2+x_2^2)\right]dx_1\,dx_2 = d$$
where d, N, h_1 and h_2 are known' (r will be an estimate for ρ).

The solution (see Appendix 4) leads to the following equation:
$$\begin{aligned}\frac{ad-bc}{N^2Z(h_1)Z(h_2)} &= r+\tfrac{1}{2}r^2h_1h_2+\tfrac{1}{6}r^3(h_1^2-1)(h_2^2-1)\\ &\quad+\tfrac{1}{24}r^4h_1(h_1^2-3)h_2(h_2^2-3)\\ &\quad+\tfrac{1}{120}r^5(h_1^4-6h_1^2+3)(h_2^4-6h_2^2+3)\\ &\quad+\tfrac{1}{720}r^6h_1(h_1^4-10h_1^2+15)h_2(h_2^4-10h_2^2+15)\\ &\quad+\tfrac{1}{5040}r^7(h_1^6-15h_1^4+45h_1^2-15)(h_2^6-15h_2^4+45h_2^2-15)\\ &\quad+\text{etc.}\end{aligned}$$
where $Z(h) = (2\pi)^{-\frac{1}{2}}\exp(-\frac{1}{2}h^2)$.

The numerical solution has to be obtained by approximating to the roots, and Newton's method * is convenient for the purpose. Tables of $L(h, k; \rho)$ (or equivalent functions) can be very helpful in suggesting good initial values, and sometimes a sufficiently close value for r may be obtained by interpolation without need to use the above equation.

* *Newton's method of approximating to the root of an equation.* Let $f(x) = 0$ be an equation from which the value of x is to be found and let b be a value near to x so that $x = b+h$ where h is small, then $f(x) = f(b+h) = f(b)+hf'(b)+$ terms involving higher powers of h by Taylor's Theorem, and since $f(x) = 0$, we have $h = -\{f(b)/f'(b)\}$ or $x = b-\{f(b)/f'(b)\}$. The chief objection the method is that there may be more than one root near the value b, but this does not hold in the application to correlation. (Cf. Approximations to rate of interest from an annuity, Todhunter's *Interest and Annuities Certain*, p. 177, formula 2.)

As an example we take the following table, drawn from vaccination statistics and relating to the Sheffield smallpox outbreak of 1887–8 (from W. R. Macdonell, *Biometrika*, I (1901), 375).

		Resistance to smallpox		
		Deaths	Recoveries	Totals
Degree of effective vaccination	Present	200	3951	4151
	Absent	274	278	552
	Totals	474	4229	4703

The observed categories may be regarded as dichotomies of the characters 'Degree of Effective Vaccination' and 'Resistance to Smallpox'. These characters were not measured quantitatively, but the absence of such measurement does not mean that there is no correlation.

We have $a = 274$, $b = 278$, $c = 200$, $d = 3951$, $N = 4703$. Hence

$$2P(h_1)-1 = \sqrt{\frac{2}{\pi}}\int_0^{h_1} e^{-\frac{1}{2}u^2}\,du = \frac{474-4229}{4703}$$

$$= -0{\cdot}7984265$$

$$\therefore \qquad h_1 = -1{\cdot}27716$$

by interpolation in Sheppard's Tables.

Similarly $$2P(h_2)-1 = \frac{552-4151}{4703} = -0{\cdot}7652561$$

$$\therefore \qquad h_2 = -1{\cdot}18833$$

From Sheppard's Tables

$$Z(h_1) = 0{\cdot}1764870; \qquad Z(h_2) = 0{\cdot}1969111$$

whence $$\frac{ad-bc}{N^2 Z(h_1) Z(h_2)} = 1{\cdot}336062$$

Dr. Macdonell gives 56 instead of 62 as the last two figures; the difference is probably due to interpolation.

Turning to the expression for r, we notice that $h_1 h_2$ is a product in the coefficients of r^2, r^4, r^6, etc., so it is well to work out its value and keep a note of it while the coefficients are being found. It is also advisable to begin the work by writing down the first six or seven powers of h_1 and h_2.

Macdonell gives the following series:

$$0{\cdot}097083r^7+0{\cdot}008170r^6+0{\cdot}119614r^5+0{\cdot}137450r^4 +0{\cdot}043352r^3+0{\cdot}758844r^2+r = 1{\cdot}336056$$

In order to obtain r we must find a value near the true one as a first approximation.

Taking $$0{\cdot}758844r^2+r-1{\cdot}336056 = 0$$

we have $$r = \frac{-1+\sqrt{\{1+4\times 1{\cdot}336\times 0{\cdot}7588\}}}{1{\cdot}5177} = 0{\cdot}8.$$

Now, this value will be in excess of the truth because we have used only two terms of the series on the left-hand side of the equation for finding r, and we may take 0·77 as a trial rate. Applying Newton's Rule, we have:

$$r = 0{\cdot}77 - \frac{\begin{array}{c}-1{\cdot}336056+(0{\cdot}77)+0{\cdot}7588(0{\cdot}77)^2+0{\cdot}0434(0{\cdot}77)^3\\ +0{\cdot}1375(0{\cdot}77)^4+0{\cdot}1196(0{\cdot}77)^5+0{\cdot}0082(0{\cdot}77)^6+0{\cdot}0971(0{\cdot}77)^7\end{array}}{\begin{array}{c}1+2(0{\cdot}77)(0{\cdot}7588)+3(0{\cdot}77)^2(0{\cdot}0434)+4(0{\cdot}77)^3(0{\cdot}1375)\\ +5(0{\cdot}77)^4(0{\cdot}1196)+6(0{\cdot}77)^5(0{\cdot}0082)+7(0{\cdot}77)^6(0{\cdot}0971)\end{array}}$$

$$= 0{\cdot}77 - \frac{0{\cdot}0022}{2{\cdot}861}$$

$$= 0{\cdot}7692.$$

In work such as this a table giving the first seven powers of the natural numbers is a help.

Systems of Type AA

5. The bivariate normal frequency surface can be taken as the generating function for more general frequency surfaces in just the same way as the normal frequency curve was used to generate Type A curves in Chapter 6. If this is done the separate (marginal) distributions of each of the two variables are found to be of Type A (as will be demonstrated below). For this reason, the corresponding frequency surfaces are said to be of Type AA.

For the present formal description, we will suppose that each variable has mean value zero and standard deviation, 1. Then in any specific application, the general theory can be applied to

[(Original Value)—(Mean)]/(Standard Deviation)

If it be supposed that the variables x_1 and x_2 have been standardized (i.e. made to have mean zero and standard deviation equal to 1) then the general equation of this surface is

$$y = \phi(x_1, x_2) + \sum_{i+j \geq 3} \sum (-1)^{i+j} \frac{A_{ij}}{i!j!} \cdot \frac{\partial^{i+j}\phi(x_1, x_2)}{\partial x_1^i \, \partial x_2^j}$$

where $\phi(x_1, x_2)$ is the standard bivariate normal frequency function already described. Various special forms can be obtained by including different sets of terms in the summations. Among these we may distinguish Type AaAa in which the marginal distributions are Type A, and Type AbAb in which they are represented by Edgeworth series.

If the right-hand side of the above equation be integrated with respect to x_2 between $-\infty$ and $+\infty$, then $\phi(x_1, x_2)$ gives the normal frequency function

$$\frac{N}{\sqrt{(2\pi)}} \exp(-\tfrac{1}{2}x_1^2) = \phi(x_1)$$

Of the remaining terms, all those with $j \geqslant 1$ will give zero since $\phi(x_1, x_2)$ and all of its partial derivatives tend to zero as x_2 tends to $+\infty$ or $-\infty$. If $j = 0$, however, we have

$$\int_{-\infty}^{\infty} \frac{\partial^i \phi(x_1, x_2)}{\partial x_1^i} \, dx_2$$

which is of form (polynomial in x_1) $\phi(x_1)$. It can be seen that the frequency function of x_1 is of Type Aa (or Ab) form. A similar argument can be applied to x_2. (If the values of the A_{ij}'s have been chosen so that the moments of the marginal distributions are correctly given then it follows that the constants in these frequency functions will be just those of the Type A (Aa or Ab) expansions for these distributions.)

Some values of the A_{ij}'s are

$$A_{30} = q_{30};\ A_{21} = q_{21};\ A_{12} = q_{12};\ A_{03} = q_{03};\ A_{40} = q_{40}-3;$$
$$A_{31} = q_{31}-3\rho;\ A_{22} = q_{22}-1-2\rho^2;\ A_{13} = q_{13}-3\rho;$$
$$A_{04} = q_{04}-3.$$

Fitting by moments, using these equations, is straightforward. For series of Type AaAa, proceeding as far as terms for which

$i+j = 4$, there are 14 values A_{ij} to be fitted. The corresponding AbAb series has extra terms with coefficients

$$A_{60} = 10A_{30}^2;\quad A_{51} = 10A_{30}A_{21};\quad A_{42} = 4A_{30}A_{12}+6A_{21}^2$$
$$A_{33} = A_{30}A_{03}+9A_{21}A_{12}$$
$$A_{24} = 4A_{03}A_{21}+6A_{12};\quad A_{15} = 10A_{03}A_{12};\quad A_{06} = 10A_{03}^2$$

These surfaces suffer the same drawbacks as the Gram-Charlier and Edgeworth curves described in Chapter 6. They can have negative ordinates; and they can be multi-modal. Further reading on these surfaces will be found in references given in Appendix 8.

6. Systems of curves obtained by supposing that the distribution of some simple function of the variable is described by the normal curve with zero mean and unit standard deviation can easily be extended to provide systems of frequency surfaces. All that is necessary, to do so, is to suppose that the joint distribution of a number of such simple functions, each of a different variable, is of multinormal form. As in other sections of this chapter, we will confine ourselves, for practical purposes, to surfaces for two variables only, but the extension to three or more is quite straightforward.

In addition to the symbols S_U, S_B introduced in Chapter 7, we will now use S_N to denote 'normal' and S_L to denote 'lognormal'. Further we will put

$$f_N(t) = t$$
$$f_L(t) = \log t$$
$$f_U(t) = \sinh^{-1} t$$
$$f_B(t) = \log\left[\frac{t}{1-t}\right]$$

and, for convenience, denote $(x-\xi)/\lambda$ by x', $(x_1-\xi_1)/\lambda_1$ by x_1' and so on.

Then we define a surface of type S_{IJ} by requiring that the joint distribution of

$$z_1 = \gamma_1+\delta_1 f_I(x_1') \quad\text{and}\quad z_2 = \gamma_2+\delta_2 f_J(x_2')$$

shall be described by the standard bivariate normal frequency surface

$$y = \frac{N}{2\pi\sqrt{(1-\rho^2)}}\exp\left[-\frac{1}{2(1-\rho^2)}\{z_1^2-2\rho z_1z_2+z_2^2\}\right]$$

Note that I, J can each be N, L, U, or B. This implies of course, that z_1 and z_2 are normal, and so x_1 is of type S_I and x_2 of type S_J.

(It is well to note, at this point, that the mere fact that z_1 and z_2

are normal does not ensure that their joint distribution is bivariate normal. Such an assumption is, however, at least a reasonable first approximation, in the absence of definite information to the contrary.)

There are ten different types of frequency surface S_{IJ}. These are S_{NN}, S_{LL}, S_{UU}, S_{BB}, S_{NL}, S_{NU}, S_{NB}, S_{LU}, S_{LB}, and S_{UB}. (Surfaces S_{IJ} and S_{JI} are essentially the same.) Of these S_{NN} is just the bivariate normal surface, which we have already discussed.

We can use our knowledge of the properties of the standard bivariate normal frequency surface to derive some properties of S_{IJ} surfaces. The conditional distribution of $z_2 = \gamma_2 + \delta_2 f_J(x_2')$, given z_1, is normal, with expected value ρz_1 and standard deviation $\sqrt{(1-\rho^2)}$. That is to say, the conditional distribution of

$$[\gamma_2 + \delta_2 f_J(x_2') - \rho\{\gamma_1 + \delta_1 f_I(x_1')\}]/\sqrt{(1-\rho^2)}, \text{ given } x_1',$$

is standard normal. This means that the conditional distribution of x_2', given x_1', is of the same type S_J as the unconditional distribution of x_2', but with γ_2 replaced by

$$[\gamma_2 - \rho\{\gamma_1 + \delta_1 f_I(x_1')\}]/\sqrt{(1-\rho^2)}$$

and δ_2 replaced by

$$\delta_2/\sqrt{(1-\rho^2)}$$

The value replacing δ_2 is the same for all x_1'. The value replacing γ_2 varies with x_1'. Since $f_I(x_1')$ increases with x_1' from $-\infty$ to $+\infty$, it is clear that (except when $\rho = 0$) the quantity replacing γ_2 will change sign at just one point in the range of variation of x_1'. If $J = B$ or U then the skewness of the conditional variation will change sign at this point.

If $J = L$, then as a consequence of the constancy of the value replacing δ_2, it follows that all the conditional distributions of x_2', given x_1', are of the same shape, but they are not of constant variance, since this also depends on the value replacing γ_2.

Expressions for the means of S_U and S_B distributions are rather cumbersome expressions, while the medians have quite simple formulae. We therefore study the median regression of x_2' on x_1'. The median of a variable having a standard normal distribution is zero. Hence the median of x_2', given x_1', satisfies the equation

$$\gamma_2 + \delta_2 f_J(x_2') = \rho[\gamma_1 + \delta_1 f_I(y_1)]$$

i.e. $$f_J(x_2') = (\rho\gamma_1 - \gamma_2)\delta_2^{-1} + \rho\delta_1\delta_2^{-1} f_I(x_1')$$

If $I = J = N$ (the bivariate normal case) then $f_J(x_2') = x_2'$ and $f_I(x_1') = x_1'$, so the regression is linear (as, indeed, we already know).

The only other cases in which the regression is linear are those in which

$$\rho\delta_1 = \delta_2; \qquad \rho\gamma_1 = \gamma_2 \quad \text{and} \quad I = J.$$

In this case the median regression is $x_2' = x_1'$. It should be noted that linear regression of x_2' on x_1' is equivalent to linear regression of $x_2 = \xi + \lambda x_2'$ on $x_1 = \xi + \lambda x_1'$.

The table below gives the median regression equations for the sixteen possible cases. (Note that there are *two* cases for each surface for which $I \neq J$, corresponding to different choices of dependent and independent variables.)

Distribution of x_2'	Distribution of x_1'	Median regression of x_2' on x_1'
S_N	S_N	$\log\theta + \phi x_1'$
S_N	S_L	$\log\theta + \phi\log x_1'$
S_L	S_N	$\theta\, e^{\phi x_1'}$
S_N	S_B	$\log\theta + \phi\log\{x_1'/(1-x')\}$
S_B	S_N	$[1+\theta^{-1}\, e^{-\phi x_1'}]^{-1}$
S_N	S_U	$\log\theta + \phi\log[x_1' + \sqrt{(x_1'^2+1)}]$
S_U	S_N	$\frac{1}{2}[\theta\, e^{\phi x_1'} - \theta^{-1}\, e^{-\phi x_1'}]$
S_L	S_L	$\theta x_1'^{\phi}$
S_L	S_B	$\theta[x_1'/(1-x_1')]^{\phi}$
S_B	S_L	$[1+\theta^{-1}x_1'^{-\phi}]^{-1}$
S_L	S_U	$\theta[x_1 + \sqrt{(x_1'^2+1)}]^{\phi}$
S_U	S_L	$\frac{1}{2}[\theta x_1'^{\phi} - \theta^{-1}x_1'^{-\phi}]$
S_B	S_B	$\theta x_1'^{\phi}[(1-x_1')^{\phi} + \theta x_1'^{\phi}]^{-1}$
S_B	S_U	$[1+\theta^{-1}\{\sqrt{(x_1'^2+1)} - x_1'\}^{\phi}]^{-1}$
S_U	S_B	$\frac{1}{2}[\theta x_1'^{2\phi} - \theta^{-1}(1-x_1')^{2\phi}]x_1'^{-\phi}(1-x_1')^{-\phi}$
S_U	S_U	$\frac{1}{2}[\theta\{x_1' + \sqrt{(x_1'^2+1)}\}^{\phi} - \theta^{-1}\{\sqrt{(x_1'^2+1)} - x_1'\}^{\phi}]$

(Note: $\theta = \exp[(\rho\gamma_1 - \gamma_2)/\delta_2]$; $\phi = \rho\delta_1/\delta_2$.)

Note that the sign of ϕ is the same as that of ρ, since δ_1 and δ_2 are assumed to be positive.

The most straightforward way of fitting a frequency surface of the kind just described is first to fit the separate distributions of x_1 and x_2 by appropriate S_I and S_J type curves respectively. Then calculate the correlation between the 'translated' values $\gamma_1 + \delta_1 f_I((x_1 - \xi_1)/\lambda_1)$ and $\gamma_2 + \delta_2 f_J((x_2 - \xi_2)/\lambda_2)$ and equate it to ρ. There will be some difficulty if the data are grouped since, in general, equal width groupings of the original values will not correspond to equal width groupings of the translated values. In such cases it is suggested that each value in a group be assigned the value of the mean of that part of the normal distribution contained in the group.

Thus, corresponding to values of x_1, between x and $(x+h)$ $(h > 0)$ z takes values between

$$z_x = \gamma_1+\delta_1 f_I\left(\frac{x-\xi_1}{\lambda_1}\right) \quad \text{and} \quad z_{x+h} = \gamma_1+\delta_1 f_I\left(\frac{x+h-\xi_1}{\lambda_1}\right)$$

Each observation in this group of values of x_1 is then assigned the z-value

$$\left[\frac{1}{\sqrt{2\pi}}\int_{z_x}^{z_{x+h}} t\, e^{-\frac{1}{2}t^2}\,dt\right]\Big/\left[\frac{1}{\sqrt{2\pi}}\int_{z_x}^{z_{x+h}} e^{-\frac{1}{2}t^2}\,dt\right]$$

$$= \left[\frac{1}{\sqrt{2\pi}}\{e^{-\frac{1}{2}z_x^2}-e^{-\frac{1}{2}z_{x+h}^2}\}\right]\Big/\left[\frac{1}{\sqrt{2\pi}}\int_{z_x}^{z_{x+h}} e^{-\frac{1}{2}t^2}\,dt\right]$$

$$= [Z(z_x)-Z(z_{x+h})]/[P(z_{x+h})-P(z_x)]$$

9. Standard errors

1. In statistical work we calculate a mean from a number of measurements, and we may be tempted to think that our work has definitely established the mean with which we are concerned. But even if the arithmetical work is correct in every detail and the measurements have been made accurately, the mean found from the statistics usually differs from the true mean of the character measured because the things measured are limited in number—because, in other words, the sample we have taken does not exactly represent the unlimited population from which it is drawn. If we toss five coins and record the number of heads, we should obtain a table like the following where we give results of 140 repetitions of the experiment.*

Number of 'heads' in trial	Number of trials in which the number of heads in previous column was recorded
5	4
4	24
3	49
2	40
1	20
0	3
	140

Now, treating this as a mere statistical problem, in which we do not know *a priori* anything about the true distribution, we may work out the mean number of 'heads' as 2·6. This does not prove that this mean and no other can arise; a second experiment might give a different result and we cannot, therefore, say from our calculation what the mean value really is. We can, however, approach the problem in another way; we can try to decide how deviations from a true mean are likely to be distributed and so form an opinion as to

* Such an experiment may, alternatively, be regarded as drawing a random sample of 140 cases from an infinite population distributed as $(\frac{1}{2}+\frac{1}{2})^5$.

how a mean calculated from an experiment or series of experiments will differ from the truth. For practical purposes we might rest content if we could say that the true mean will not differ from the calculated mean by more than a small quantity (ϵ) once in a hundred trials. Before we go into the measures actually used, let us consider some of the points in a preliminary way.

All our statistical experience makes us feel sure that an experiment based on 1000 cases must be more reliable than a similar experiment based on 50 cases, so we anticipate that ϵ, the difference between the true and the calculated mean, will depend in some way on the number of cases. Again, a distribution that spreads widely gives the mean more opportunity to deviate than a distribution that is concentrated; so that we may also anticipate that ϵ will depend on the spread of the distribution, that is, on its standard deviation.

2. These remarks apply to all statistical measures. The measures are inexact and only approximate to the truth, but we can say that it is highly probable that they do not differ by more than a certain amount from the result which would be obtained if we could deal with an unlimited number of facts. In our discussion we have spoken of means, but every other measure is subject to the same general considerations and we must, therefore, consider what sort of value may be assigned to the small error ϵ for means, standard deviations, coefficients of correlation and other measures. We have anticipated that this error will depend on a standard deviation; and this sometimes leads to a little confusion because we must make up our minds as to the distribution to which the standard deviation refers. Let us imagine that we have worked out a coefficient of correlation for 100 pairs of, say, ages at marriage of husband and wife. Then we work out a second coefficient of correlation for another hundred pairs and so on until we have a large number of these results. The coefficients will fall into a distribution like one of the frequency distributions we discussed in earlier chapters and it is the standard deviation of that distribution from its mean with which we are concerned. Similarly, we can repeat the coin-tossing experiment of § 1 over and over again and calculate the mean from each experiment. We may obtain any value for the mean between 5 and 0 heads; these extremes will only arise in the most unlikely case when every trial gave 5 heads or every trial gave no head. If we repeat the experiment sufficiently we shall form an idea of the way the means are distributed: we shall reach a frequency distribution of means having its own mean, standard deviation, etc., and we must not confuse it with the frequency distributions such as that in the table in § 1 from which the means were calculated. It will help to avoid confusion of ideas if we speak of 'standard error' when we are referring to the frequency distribution of a statistical measure (such as a mean, coefficient of correlation, etc.) instead of speaking of 'standard deviation'. The standard error is, then, the standard deviation of the frequency distribution of the particular measure we are examining.

3. With this introduction we may now consider the simplest of the bell-shaped frequency curves, namely, the normal curve of error, and see what conclusions we may draw if the distribution of a statistical measure takes that form. It has, in fact, been shown to be the form that the distribution of statistical measures usually tends to assume when the number of cases in the sample is large. Thus, even if the distribution in § 1 had been skew instead of symmetrical, the distribution of the means would have been more nearly of the form of the normal curve than the skew distributions from which they were obtained. By reference to the tables of this curve we see that the area corresponding to the standard deviation is about two-thirds of the whole area, while the area corresponding to twice the standard deviation is 0·9545 of the whole area.

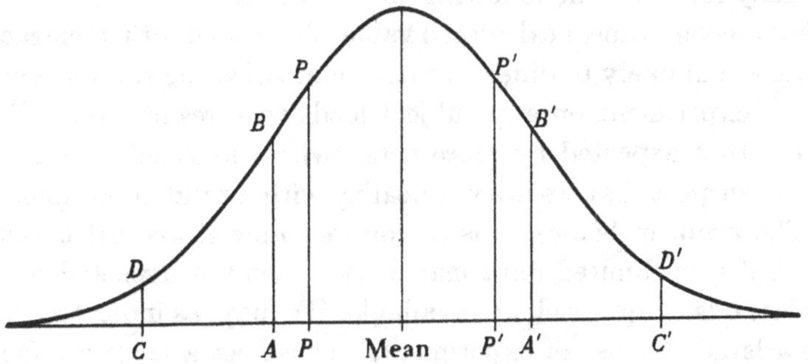

In other words, if the distribution takes this form we can say that an error of more than twice the standard error will occur 9 times in 200 trials and is, therefore, unlikely to have arisen in the particular case with which we are dealing. The diagram will help the reader to follow this argument. The area between AB which is at a distance equal to the standard deviation from the mean one side, and $A'B'$ at the same distance the other side, of the mean, is approximately two-thirds of the whole area. The lines CD and $C'D'$, which are twice as far from the mean, include nearly the whole curve; the pieces beyond those lines are tails which must be of relatively small dimensions.

4. It was formerly the custom to use another function known as the probable error, which is 0·67449 times the standard error. The probable error gives that value of x (say p) which divides the part of the normal curve representing positive errors into two equal portions; it is therefore given by

$$\int_0^p \frac{1}{\sqrt{(2\pi)}} e^{-\frac{1}{2}x^2} dx = 0{\cdot}25$$

where the whole area of the curve (positive *plus* negative deviations) is unity. In order to find p in terms of the standard deviation, we have, therefore, to obtain the value of x, corresponding to $P(x) = 0{\cdot}75$ in *Biometrika Tables*, Table 2 or the short table in Appendix 9.

This can be done by interpolating inversely and p is thus found to be 0·67449 approximately. The mean, or rather the vertical through the mean, divides the whole distribution into two equal parts: the probable error divides it into fourths and gives what Galton called the quartiles. The position is shown with the letters P and P' in the diagram; thrice the probable error includes about the same area as twice the standard error.

We may set down the following general rules:

(1) the true value and a calculated value of a mean or other characteristic are unlikely to differ by more than twice the standard error,

(2) if an experiment on any subject leads to a result which differs from that expected by more than twice the standard error we must suspect that we are not dealing with a random sample.

5. The problem before us is to consider how statistical measures calculated from limited data may vary about the expected values. Two methods of approach are available. We may, as indicated in § 2, make a large number of experiments—or collect a large number of samples—and calculate the statistical measure in question for each of them. The procedure would generally be much too laborious, and we take, therefore, the second line of approach. Algebraic analysis based on the theory of probability enables us to determine the standard error that we should find in the limit if the sampling process were repeated indefinitely so that all possible samples were included in their expected proportions. We can often go further and determine the actual curve to which the frequency distribution of a particular statistical measure will tend as the number of samples is increased.

We may now take a simple illustration and find the standard error of the frequency, say n, with which an event will happen in m independent trials where p is the probability of it happening and q of it failing. The probability of n being equal to $m, m-1, \ldots, 2, 1, 0$ is given by the terms of the binomial expansion. Taking moments about the point represented by p^m, the first moment is

$$mp^{m-1}q+m(m-1)p^{m-2}q^2+\ldots+mq^m = mq(p+q)^{m-1} = mq.$$

The second moment about the same point is

$$\begin{aligned} &mp^{m-1}q+2m(m-1)p^{m-2}q^2+\tfrac{3}{2}m(m-1)(m-2)p^{m-3}q^3+\ldots+m^2q^m \\ &\quad = mp^{m-1}q+m(m-1)p^{m-2}q^2+\frac{m(m-1)(m-2)}{2!}p^{m-3}q^3 \\ &\qquad +\ldots+mq^m+m(m-1)p^{m-2}q^2+m(m-1)(m-2)p^{m-3}q^3 \\ &\qquad\qquad +\ldots+m(m-1)q^m \\ &\quad = mq+m(m-1)q^2 \end{aligned}$$

The second moment about the mean is, therefore,

$$mq+m(m-1)q^2-m^2q^2 = mpq$$

$$\therefore \qquad \text{the standard error} = \sqrt{\mu_2} = \sqrt{(mpq)}$$

That is to say, if we repeatedly make m independent trials the observed frequency of occurrence, n, will vary about the expected value mp with a standard error of $\sqrt{(mpq)}$.

6. We may now apply this result to a few examples:

(*a*) It has been remarked that the number of male children born is to the number of female children born as 1050 : 1000; in other words, the probability of a child being male is 1050/2050. If 51 350 out of 100 000 children proved to be males in a certain community, would it be safe to base on the statistics any theory connected with the variation from the usual probability? The expected result is 51 220, and the standard error is

$$\sqrt{\left(100\,000 \cdot \frac{1050}{2050} \cdot \frac{1000}{2050}\right)} = \pm 158{\cdot}07$$

The difference between the actual case and the expected result was 130, and as this is less than the standard error, no definite conclusion can be based on the divergence from the result.

(*b*) If the number of cases had been 10 000 000, and the actual number 5 135 000, then the standard error being 1580·7 and the actual difference 13 000, it would have been sufficient evidence for the conclusion that the ratio 1050 : 1000 did not fit the particular case.

(*c*) It will be noticed that if m remains constant, then $\sqrt{(mpq)}$ has its largest numerical value when $p = q = \frac{1}{2}$, which shows that an insurance office will generally find that if it has two classes of equal size, and one is subject to a higher rate of mortality than the other, the former will have the larger actual deviations from the expected number of claims, because the probability of dying in a year only reaches the value $\frac{1}{2}$ near the end of the mortality table.

7. We may now consider a frequency distribution divided into k groups such that the proportion of cases in the sth group is p_s and, clearly, $p_1+p_2+\ldots+p_s+\ldots+p_k = 1$. If we take a case at random from the distribution, the chance that it comes from the sth group is p_s and the chance that it comes from some other group is $q_s = 1-p_s$. Let us suppose that m cases are taken at random and that n_s of them fall in the sth group. Then, though the expected value of n_s is mp_s, this frequency may assume values m, $m-1$, ..., 2, 1, 0 with probabilities given by the terms of the binomial $(p_s+q_s)^m$ and the standard error of n_s will be

$$\sigma_{n_s} = \sqrt{\{mp_s(1-p_s)\}} \qquad \ldots\ldots(1)$$

If, in practice, we do not know the exact form of the frequency distribution from which the sample has been taken, we may approximate to the standard error by putting $p_s = n_s/m$, the observed proportion in the sth group of the sample. Hence, we have, approximately,

$$\sigma_{n_s} = \sqrt{\{n_s(1-n_s/m)\}} \qquad \ldots\ldots(2)$$

and if m is large, compared with n_s, this is very nearly equal to $\sqrt{n_s}$.

8. As the total of all the frequencies n_1, n_2, ... n_k is m, it follows that, if in a particular sample n_s is much greater than mp_s, the other frequencies must on the average be too small and this shows that the errors between the groups are correlated. The next point to be investigated is the amount of *the correlation between deviations in the frequencies of the s-th and t-th groups.*

The deviation, δn_s, of n_s from its expected value is n_s-mp_s.

As we are considering the relation between deviations in n_s and n_t

we may conveniently class together all the remaining $k-2$ frequency groups into a single remainder group, say, n_R. Then

$$n_s+n_t+n_R = m$$
$$p_s+p_t+p_R = 1$$
$$\delta n_s+\delta n_t+\delta n_R = 0$$

and $$(\delta n_s+\delta n_t)^2 = (-\delta n_R)^2$$

or $$\delta n_s \delta n_t = \tfrac{1}{2}(\delta n_R^2-\delta n_s^2-\delta n_t^2)$$

If we now imagine that a very large number, N, of random samples is taken and the expressions on both sides of the last equation are summed and divided by their number, N, then

$$\frac{1}{N}S(\delta n_s \delta n_t) = \tfrac{1}{2}\left\{\frac{1}{N}S(\delta n_R^2)-\frac{1}{N}S(\delta n_s^2)-\frac{1}{N}S(\delta n_t^2)\right\}$$

The expressions on the right-hand side represent, in the limit, the squared standard errors of the group frequencies, given in equation (1) above. Hence in the limit

$$\begin{aligned}\frac{1}{N}S(\delta n_s \delta n_t) &= \tfrac{1}{2}m\{p_R(1-p_R)-p_s(1-p_s)-p_t(1-p_t)\}\\ &= \tfrac{1}{2}m\{(1-p_s-p_t)(p_s+p_t)-p_s(1-p_s)-p_t(1-p_t)\}\\ &= -mp_s p_t \qquad \text{......(3)}\end{aligned}$$

But the correlation between n_s and n_t is

$$\begin{aligned}r_{n_s n_t} &= \frac{\text{Limit of } \frac{1}{N}S(\delta n_s \delta n_t)}{\sigma_{n_s}\sigma_{n_t}}\\ &= -\frac{p_s p_t}{\sqrt{\{p_s(1-p_s)p_t(1-p_t)\}}}\\ &= -\sqrt{\left\{\frac{p_s p_t}{(1-p_s)(1-p_t)}\right\}} \qquad \text{......(4)}\end{aligned}$$

We may again approximate to this expression by substituting for p_s and p_t the proportionate frequencies, n_s/m and n_t/m, of the sample.

9. *To find the standard error of the mean of a sample of m observations.*

Let us again assume a frequency distribution divided into k groups where x_s is the value of the variable quantity x associated with the sth group. For the reasons already explained in the earlier sections of this chapter we must distinguish between (i) the mean of the population represented by the frequency distribution, namely

$$\overline{X} = \sum(p_s x_s)$$

where $\sum$ indicates summation for all the k groups, and (ii) the mean calculated from a particular sample of m cases drawn at random from this population, namely

$$\bar{x} = \sum\left(\frac{n_s}{m}\,.\,x_s\right)$$

The standard error of $\bar{x}$, say $\sigma_{\bar{x}}$, will provide a measure of the extent to which the mean of the sample may differ from the mean of the population. The value of $\sigma_{\bar{x}}$ may be found by using the results (1) and (3) of the preceding sections.

Using a similar notation, we have

$$\begin{aligned} \delta\bar{x} &= x - \overline{X} \\ &= \frac{1}{m}\sum(n_s - mp_s)x_s \\ &= \frac{1}{m}\sum(\delta n_s)x_s \end{aligned}$$

As the expected value of δn_s is zero, the expected value of $\delta\bar{x}$ is zero, or the mean value found from repeated sampling of the mean of the sample is the same as the mean of the population.

Squaring both sides of the last equation above, we have

$$(\delta x)^2 = \frac{1}{m^2}\{\sum(\delta n_s^2 x_s^2) + 2\sum{}'(\delta n_s \delta n_t x_s x_t)\}$$

where $\sum{}'$ indicates summation for all parts of values of s and t for which s is not equal to t.

If we now assume a large number, N, of samples to have been taken and the corresponding values of $(\delta\bar{x})^2$ summed and the result divided by N, we obtain

$$\frac{1}{N}S(\delta\bar{x})^2 = \frac{1}{m^2}\left\{\sum\left(x_s^2 \frac{1}{N}S(\delta n_s^2)\right) + 2\sum{}'\left(x_s x_t \frac{1}{N}S(\delta n_s \delta n_t)\right)\right\}$$

where S denotes the summation in respect of the N samples. The left-hand side of this equation is the squared standard error of the mean of the sample, or $\sigma_{\bar{x}}^2$. On the right-hand side $(1/N)S(\delta n_s^2)$ is the $\sigma_{n_s}^2$ of equation (1) and $(1/N)S(\delta n_s \delta n_t)$ is given in equation (3). Hence

$$\begin{aligned} \sigma_{\bar{x}}^2 &= \frac{1}{m^2}\{\sum[x_s mp_s(1-p_s)] - 2\sum{}'(x_s x_t mp_s p_t)\} \\ &= \frac{1}{m}\sum(p_s x_s^2) - \frac{1}{m}\{\sum(p_s x_s)\}^2 \\ &= \frac{1}{m}(\mu_2' - \overline{X}^2) \end{aligned}$$

where μ_2' is the second moment, about the origin for x, of the distribution of the population. But $\mu_2'-\overline{X}^2 = \sigma_x^2$, therefore

$$\sigma_{\bar{x}} = \sigma_x/\sqrt{m} \qquad \text{......(5)}$$

We thus find that the standard error of the mean is the ratio of the standard deviation in the population to the square root of the size of the sample.

10. This last result is of considerable use in statistical work. A large number of cases is recorded and the mean used to compare the particular experiment with another of a like kind. Is an actual difference between the means due to some cause other than random sampling? A practical application would be the comparison of the average profit from various classes of business for a number of years. The standard error of the profits in the various years would be obtained by taking the square root of the second moment about the mean and dividing it by the square root of the number of years; the quotient would give $\sigma_{\bar{x}}$ of (5). It is only by using the standard errors (or probable errors deduced from them) that we could say definitely whether a lower average profit in a certain part of the business was due to chance or to some causes requiring removal.

11. In § 5 of this chapter it was mentioned that we can often determine the actual curve to which the frequency distribution of a statistical measure tends. We saw, in Chapter 4, that β_1 and β_2 could, with the mean, be used to fix the frequency curve if it is of the Pearson family of curves, and it follows that if we can find β_1 and β_2 for the frequency distribution of a statistical measure we shall have gone a long way towards fixing the form of the curve. If we write $\beta_1(x)$ and $\beta_2(x)$ as the moment ratios for the population distribution of x and $\beta_1(\bar{x})$ and $\beta_2(\bar{x})$ for the distribution of $\bar{x}$ (the sample mean) in repeated samples of size m, then it can be shown (see R. Henderson, *J. Inst. Actu.* XLI, 429) that

$$\left.\begin{aligned} \beta_1(\bar{x}) &= \beta_1(x)/m \\ \beta_2(\bar{x}) &= 3+\{\beta_2(x)-3\}/m \end{aligned}\right\} \qquad \text{......(6)}$$

Thus if the distribution of x is represented by the normal curve for which $\beta_1(x) = 0$ and $\beta_2(x) = 3$, it is seen that

$$\beta_1(\bar{x}) = 0 \quad \text{and} \quad \beta_2(\bar{x}) = 3$$

and the distribution of $\bar{x}$ is also normal. Even if the distribution of x is not normal, it follows from equations (5) that $\beta_1(\bar{x})$ approximates to zero and $\beta_2(\bar{x})$ approximates to 3 if m is not too small.

12. The standard error of a standard deviation may be taken as $(\sigma_x/2)\sqrt{\{(\beta_2(x)-1)/m\}}$ for large samples and when the distribution of the population approximates to the normal curve of error (when $\beta_2(x) = 3$) the standard error becomes $\sigma_x/\sqrt{(2m)}$.

Another standard error which is often useful relates to the difference between two percentages or proportions. Thus if we make m_1 trials and the event happens n_1 times and in an independent m_2 trials we find n_2 happenings, in what circumstances can we conclude that $p_1 = p_2$ where the sample estimate of p_1 is n_1/m_1 and of p_2 is n_2/m_2? The solution might be useful when two rates of mortality, withdrawal or sickness are being compared.

If $p_1 = p_2 = p$, say, the standard error of the difference $n_1/m_1 - n_2/m_2$ is $\sqrt{\{p(1-p)(1/m_1+1/m_2)\}}$. We do not really know p, the underlying proportion to which the p_1 and p_2 of our experiments approximate, but on the hypothesis that there is a common value we may makes an estimate of it from

$$(n_1+n_2)/(m_1+m_2).$$

This leads to a standard error of

$$\sqrt{\left\{\frac{n_1+n_2}{m_1+m_2}\left(1-\frac{n_1+n_2}{m_1+m_2}\right)\left(\frac{1}{m_1}+\frac{1}{m_2}\right)\right\}}$$

As an example we may take (1) 1000 cases with 22 withdrawals giving a rate of withdrawal of 0·0220 and (2) 600 cases with 19 withdrawals giving a rate of 0·0317. Is the difference 0·0097 significant? The combination of the two experiences gives 41/1600 or 0·0256 as the rate of withdrawal. The standard error by the formula last given is 0·0082. The difference is not significant. If however the numbers had all been three times greater, the standard error would have been 0·005 and it would require little additional evidence to satisfy us that the difference is significant.

13. In similar ways it is possible to find the standard errors of the moments and constants, but this leads to the more theoretical parts of the subject with which it is inadvisable to deal in a book of this character. It is, however, necessary to call attention to certain standard errors owing to their importance in statistical work. We first consider the sample correlation coefficient.

As in the case of the mean, it will help to avoid confusion if we use a symbol, ρ, for the correlation coefficient in the population itself different from the symbol, r, for the coefficient calculated from a particular sample of m pairs of observations. From one sample to another r will vary about ρ and it has been shown that the standard error of r is, for large samples,*

$$\sigma_r = (1-\rho^2)/\sqrt{(m-1)} \text{ approximately} \qquad \ldots\ldots(7)$$

If we do not know ρ we use r as an approximation to it.

This formula applies to the 'product-moment correlation' calculated from the sample value of $p_{11}(p_{02}p_{20})^{-\frac{1}{2}}$ in the notation introduced in Chapter 8. For an estimate from a 2×2 table (see Chapter 8) the standard deviation may be expected to be considerably greater (two, or even three, times as big).

R. A. Fisher suggested (*Metron*, I, 1921) an useful transformation to

$$\tfrac{1}{2}\{\log_e(1+r)-\log_e(1-r)\}$$

which is distributed approximately normally with a standard error of

$$1/\sqrt{(m-3)}$$

whatever the value of ρ. This approximation gives sufficiently accurate results for most practical purposes. More precise values can be obtained, using F. N. David's *Tables of the Correlation Coefficient*.

When the distribution is really normal, the standard error of the moment estimator of $\sqrt{\beta_1}$ is approximately $\sqrt{6/N}$ and that of the moment estimator of β_2, approximately $\sqrt{24/N}$, where N is the sample size.

14. It will be noticed that, as we anticipated, all the expressions for the standard errors contain the square root of the number of cases in the denominator. We anticipated in the first paragraphs of this chapter that the standard error would decrease as the number of cases increased and we can now say that in each of the cases discussed the standard error varies inversely with the square root of the number of cases. The student should make it a rule to work out standard errors and he will find that much labour can be saved by using tables, that have been published in *Biometrika Tables* and other sources. Often complete distributions are given, and not only standard errors.

The object in calculating standard errors is to prevent ourselves from reading too much from the means or other measures we have

* When m is large $\sqrt{m}$ can be used for $\sqrt{(m-1)}$ here and in similar formulae.

calculated, but we must not run to the opposite extreme and rely more on a standard error than the theory justifies. Thus, at certain points, our theory has assumed that the characteristics are distributed in a form approximating to a normal curve of error, and a good deal of evidence has been produced showing that this is a reasonable assumption for many characteristics when the number of cases exceeds 30, or for some characteristics with even smaller numbers. The assumptions imply that plus and minus errors are equally probable, but it would not be right to assert that the means of a sample of a **J**-shaped distribution are equally likely to fall above and below the true mean within twice the standard error, and formulae (6) above help to indicate this limitation.

15. We may now refer briefly to some practical points in 'sampling'. The essence of sampling is that we form an opinion of the whole by examining a sample of it, and error may arise (1) owing to bias in making up the sample of (2) owing to the particular sample giving a wide deviation from the whole because it is based on a small number of cases.

It is, usually, not difficult to guard against bias in actuarial or sociological practice. For instance, if we require to estimate the mortality of lives assured we might collect information merely in respect of persons whose names begin with A. This would give fewer cases, but there is no reason to suspect that such lives differ from those whose names begin with the other letters of the alphabet. The selection of a particular letter might, however, lead to suspicion if it could introduce a question of race in a mixed community, e.g. in Alsace-Lorraine; if we worked with people whose names begin with W we should exclude those of French extraction but include those of German extraction. An alternative is to take one case in, say, each hundred, e.g. the mortality of lives assured could be investigated by examining from the registers of the insurance offices every hundredth case.

Sampling of this kind is useful in social investigations where we may, perhaps, want to examine the home conditions of school children and cannot hope to get from every home particulars of the health, occupation or habits of the residents. We might, however, be able to make an exhaustive examination of 2000 or 3000 cases. With a free hand it is not difficult to obtain a random sample, and a little thought and common sense is all that is required.

The other risk of error lies in the fact that we have only a small sample, and it is here that the subject is connected with that of 'standard errors'. If we may assume that the sample is chosen at random and, though not of itself small, is small compared with the population from which it is drawn, then we can follow the methods indicated in the earlier part of this chapter.

16. Special circumstances, however, arise in some experiments, and one type of case may be specially mentioned. It is frequently necessary to test the comparative yields of different varieties of the same plant. The trouble in such a case is that plots placed far apart even in a small field produce widely different results, but small adjacent plots resemble each other. In order to make a fair comparison we ought, therefore, to have a number of pairs of adjacent plots. The comparison is made between a number of pairs and we are concerned with the differences between these pairs and must work out

$$\sigma^2 = \frac{S(y-x)^2}{m}$$

where m is the number of 'pairs', and x and y are the corresponding members of a pair measured from their means.

It is important to distinguish this sort of case, where the pair formed from adjacent plots is the unit, from the different case where we draw a sample of m_1 observations from one record with a standard deviation of σ_1 and a second sample of m_2 from another record in which the standard deviation is σ_2. In this case the standard deviation of the difference between the two means is given by

$$\sigma^2 = \frac{\sigma_1^2}{m_1} + \frac{\sigma_2^2}{m_2}$$

This assumes that there is no correlation between the variables, but in the 'pairs' problem we have arranged 'pairs' because we expect correlation. Algebraically the correlation is indicated by the xy term of $S(y-x)^2 = S(y^2-2xy+x^2)$.

It will be appreciated from what has been written elsewhere in this chapter that it is assumed that the samples are sufficiently large to justify the assumption that the σ's calculated from the samples can be treated as the standard deviations of the population.

The use of the wrong formula may lead to erroneous conclusions: the actual difference between the means may be 30, the standard error by $\sigma^2 = S(y-x)^2/m$ may be 6, and by $\sigma^2 = (\sigma_1^2/m_1)+(\sigma_2^2/m_2)$

may be 12. Judged by the former the difference is almost certainly significant; judged by the latter it is doubtful.

The kind of problem indicated might arise whenever it is necessary to compare the results of alternative methods in changing conditions, and the theory which was worked out primarily to test may prove valuable elsewhere.

10. Tests of goodness of fit

1. When the values of ordinates and areas were calculated in the examples of the various types of frequency curves, no systematic attempt was made to test the graduations in order to ascertain whether the results obtained were reasonable. Actuaries have generally been in the habit of imposing on the graduated values of any table on which they may have been working, rough checks which have amounted to a comparison of the totals in various groups and an inspection of the changes of sign in the differences between the graduated and ungraduated figures. The problem of the goodness of fit needs, however, more accurate treatment; for inspection, even when aided by the calculation of a standard error for each group, can only tell that certain differences are large, and if the standard error be exceeded in two or three cases, it is impossible to say whether these impair the general quality of the rest of the graduation. A test is required which will give some measure of the disagreement as judged by the whole graduation.

2. Now, if there be N observations distributed in n groups, the numbers in the groups being $m'_1, m'_2, \ldots, m'_n$, we have to find a criterion to enable us to decide when the series $m_1, m_2, \ldots, m_n$ will be a legitimate graduation. We may clearly take a legitimate graduation to be one in which the observed values (m') do not differ from the theoretical (m) by more than the deviations that would be expected in random sampling. What we require to know is not the probability that the particular series of m''s will occur if the m's represent the theory, but the probability that the m''s, or an *equally likely or less likely* series, will arise. To appreciate the difficulties of the problem we may consider the simplest case, that of a coin-tossing experiment, and suppose that a coin has been tossed six times and come down four heads and two tails. The 'graduation' we make is three heads and three tails, and to test it we require to find the probability of obtaining a result as unlikely, or more unlikely than the observed one. The probability is the same as that of getting any one of the following results:

6 heads and 0 tails
5 heads and 1 tail

4 heads and 2 tails

2 heads and 4 tails

1 head and 5 tails

0 heads and 6 tails

It is impossible to calculate such probabilities directly, even when the simple probabilities leading to the deviations are known, in any but the easiest cases; but when we do not know the simple probabilities, or the case is a complicated one, a further difficulty is introduced owing to our inability to tell from *a priori* reasoning which of the possible cases are more or less likely than that which has actually arisen. It would, for instance, be difficult to say, without a large amount of arithmetical work, when 20 dice were being thrown, whether the probability of getting ten 'sixes' or more was greater than that of getting two 'sixes' or fewer; but this is an extremely simple case compared with the general proposition in which deviations over a series of numbers have to be considered.

3. If it is assumed in any measurement on one subject that the deviations from the mean take the form of the 'normal curve of error', and it is required to estimate the chance of obtaining deviations greater than a certain value (t, say), it will be necessary to sum all values of the normal curve beyond t on each side of the mean, i.e. we must take

$$\frac{N}{\sigma\sqrt{(2\pi)}}\left[\int_{-\infty}^{-t} e^{-\frac{1}{2}(x/\sigma)^2}\,dx+\int_{t}^{\infty} e^{-\frac{1}{2}(x/\sigma)^2}\,dx\right]=\frac{2N}{\sqrt{(2\pi)}}\int_{t/\sigma}^{\infty} e^{-\frac{1}{2}x^2}\,dx$$

where σ is the standard deviation and divide the result by the area of the whole curve, i.e. N. Assuming that there are two measurements instead of one (the exposed to risk, for instance, at two ages), the deviations are as it were, in two directions instead of one, and it is necessary to take an expression with two variables instead of one. The expression analogous to the normal curve is the correlation surface

$$z = z_0 e^{-\frac{1}{2}\left\{\frac{x^2}{\sigma_1^2}-\frac{2xy\rho}{\sigma_1\sigma_2}+\frac{y^2}{\sigma}\right\}/(1-\rho^2)}$$

with which we have already dealt. The integrations must be done over a region $(x/\sigma_1)^2-2\rho(x/\sigma_1)(y/\sigma_2)+(y/\sigma_2)^2 > t^2$ (since the surface height is a decreasing function of $(x/\sigma_1)^2-2\rho(x/\sigma_1)(y/\sigma_2)+(y/\sigma_2)^2$)

and compared with the total. It is natural to measure departure from the mean value by the quantity

$$\chi^2 = (1-\rho^2)^{-1}\left[\left(\frac{x}{\sigma_1}\right)^2 - 2\rho\left(\frac{x}{\sigma_1}\right)\left(\frac{y}{\sigma_2}\right) + \left(\frac{y}{\sigma_2}\right)^2\right]$$

which appears in the exponent of the expression for z, or by the corresponding quantity for ν variables, and to seek the probability that it exceeds a fixed quantity, t^2 say. (This is equivalent to finding the probability that the ordinate z is less than $z_0 e^{-\frac{1}{2}t^2}$.) If there are ν measurements it becomes necessary to deal with a function of ν variables, and this will give the reader a slight idea of the problem from the mathematical point of view. It has been shown * that the result

$$P = \frac{\int_t^\infty e^{-\frac{1}{2}x^2} x^{\nu-1}\,dx}{\int_0^\infty e^{-\frac{1}{2}x^2} x^{\nu-1}\,dx} \dagger$$

is reached. The number ν is called the 'degrees of freedom' of the quantity χ^2.

4. Before a measure of the probability P can be obtained a value for χ must be found from the statistics of the particular graduation, and in the paper to which reference has already been made its value is shown to be such that we can take, with good approximation

$$\chi^2 = S\left\{\frac{(m_r - m_r')^2}{m_r}\right\}$$

The appropriate value of ν (the degrees of freedom) is n *minus* the number of separate parameters estimated from the data. Thus if only N is 'estimated' (i.e. the frequency curve or surface is given *a priori*) then $\nu = n-1$, but if the mean and standard deviation are also estimated (as, for example, in fitting a normal frequency curve), then $\nu = n-3$. As further examples (*a*) if a bivariate normal frequency surface is fitted using two means, two standard deviations and a correlation coefficient, then $\nu = n-6$, and (*b*) if a S_{UU} surface is fitted with, apart from N, each marginal distribution requiring four

*Originally by K. Pearson: 'On the criterion that a given system of deviations from the probable in the case of a correlated system of variables is such that it can be reasonably supposed to have arisen from Random Sampling', *Phil. Mag.*, July 1900. A short proof has been given by H. E. Soper in *Frequency Arrays*.

† A table of P for all values of ν from 1 to 70 for P up to about 0·99999 is given in *Biometrika Tables*. An abridged table is given in Appendix 9.

parameters to be fitted, and the correlation coefficient also being used, then $\nu = n-10$. (See also section 8.)

It is natural, almost necessary, to use the square of the difference in order that negative differences may, equally with positive differences, increase the improbability of the system, while a ratio is required to bring into account the size of the group, for an error of 15 in a group of 20 would be very large, but in a group of 1000 would be negligible.

5. The practical aspects of this test of goodness of fit and its application may now be dealt with.

(1) If the facts representing the graduated and ungraduated figures are only available in groups, then the value of the probability by the test will, as a rule, be lower as the number of groups is increased. This practical point should be borne in mind as it sometimes happens that graduations are tested in groups of, say, five years of age; but the graduated figures for individual ages are then used unreservedly, though, strictly speaking, they may be no better than interpolated values.

(2) The test assumes a distribution, and would not be applicable if the numbers were a series of ordinates, though the application of the test would probably give a fair idea of the goodness of fit if a large number of ordinates had been given in the series.

(3) The tails of the experience will be very small and never fit exactly. We ought to take our final theoretical groups to cover as much of the tail area as amounts to at least a unit of frequency in such cases.

(4) If the number of observations be multiplied by t, say, and the deviations are also multiplied by t, then the value of χ^2 will be multiplied by the same figure, and the test will show that the fit is worse. This may seem strange at first, but a little consideration will show that it is reasonable. As a large number of cases will give smoother series than a small number, it follows that if two results are proportionally the same in two examples having the same theoretical distribution but different total frequencies, the one with greater frequency is less probable than the one with less frequency. The probability of a result as bad as, or worse than, three heads and one tail in coin-tossing (two heads and two tails being the theoretical result) is 0·625; but the probability of a result as bad as, or worse than, $3\times 2 = 6$ heads and $1\times 2 = 2$ tails is 0·289. It follows that if

a distribution is based on, say, 103 480 cases and the figures are reduced to a total of 1000 to show the distribution of the cases, then a graduation tested as if 1000 were the total frequency will give the impression that the graduation is far closer than it really is.

(5) I have found, in applying the test, that when the numbers dealt with are very large the probability is often small, even though the curve appears to fit the statistics very closely. The explanation may be that the statistics with which we deal in practice nearly always contain a certain amount of extraneous matter, and the heterogeneity is concealed in a small experience by the roughness of the data. The increase in the number of cases observed removes the roughness, but the heterogeneity remains. The meaning, from the curve-fitting point of view, is that the experience is really made up of more than one frequency curve; but a certain curve, approximating to the one calculated, predominates. Another possible explanation is that our solution of the problem depends on the assumption of a mathematical expression which does not give exactly the distribution of deviations and when we deal with a large experience the approximate nature of the assumptions is revealed.

(6) What is the actual value of P at which a good fit ends and a bad one begins? It is impossible to fix such a value. We have merely a measure of probability for the whole table, and if the odds against the graduation are 20 or 30 to 1 the result is unsatisfactory; if they are ten to one the graduation is not unreasonable, but the exact value when a result must be discarded cannot be given. As, however, it is clearly impossible to imagine any test which can fix an absolutely definite standard, there is no reason for objecting to the particular method because if fails to do so.

(7) It is sometimes thought that the introduction of additional constants must necessarily improve the fit of a curve. It may do so in some cases, but it is quite possible to take a curve with ten constants and find it gives a worse result than another having only three. Besides this, there is the possibility of undergraduation; we must not expect to reach a very high value for P, e.g. 0·95. If we make an experiment in coin-tossing, it is unlikely that a single experiment will give a distribution very close to the theoretical. If therefore we are estimating the probability of getting that result or worse, we shall only rarely get a very high or a very small value for that probability. We shall do so occasionally, but we must not

expect it and it is wise to look for explanations when any graduation gives a very high or very low value of P.

(8) It may sometimes be advisable to use a curve giving a worse agreement than another for simplicity, or for reasons such as those which prompt actuaries to employ Makeham's hypothesis.

6. In a paper 'On the Comparative Reserves of Life Assurance Companies, etc.' (*J. Inst. Actu.* XXXVII, 458–9), George King remarked that it is permissible to use the H^M Model Office for the O^M; and it will be interesting to apply the formulae given above to see what is the probability of the O^M distribution if the H^M be taken as the theoretical distribution.

In the table following there are ten groups, and $\chi^2 = 1{\cdot}79$, and *Biometrika Tables* give $P = 0{\cdot}99944$ and $0{\cdot}99147$ when $\chi^2 = 1$

Central age in group	Policies issued arranged in age-groups		$O^M - H^M$		(Square of $O^M - H^M$)/H^M
	H^M	O^M	+	−	
20	6·97	7·30	0·33	...	0·02
25	17·75	20·45	2·70	...	0·41
30	21·04	23·11	2·07	...	0·20
35	18·41	18·40	...	0·01	0·00
40	13·82	13·05	...	0·77	0·04
45	9·45	8·44	...	1·01	0·11
50	6·23	5·07	...	1·16	0·22
55	3·51	2·58	...	0·93	0·25
60	1·97	1·20	...	0·77	0·30
65	0·85	0·40	...	0·45	0·24
...	100·00	100·00	5·10	5·10	$\chi^2 = 1{\cdot}79$

and 2 respectively (9 degrees of freedom). But it is not enough to test for 100 new policies. 950 would reduce the probability to about 0·05, which means that in only one case out of twenty would random sampling lead to a system of deviations from the H^M as great as that shown by the O^M. This result will remind the student of the great danger of dealing with percentages without considering the actual number of cases investigated. King's other table, which is of greater importance in his work (policies according to attained age), shows a much closer agreement, as $P = 0{\cdot}831051$ for 10 000 cases.

7. We will now revert to § 2 of this chapter where in stating the problem it was said that the N observations were distributed in

n groups. As we have only N observations to distribute we can only choose $n-1$ groups, for having fixed those groups the last one is necessarily fixed; freedom of choice is restricted to this extent, and in any problem where the method is used the number of groups where freedom of choice is possible must be borne in mind. This is implied in the proofs leading up to the formulae which have been given. Now, following on this argument the reader may ask whether it is fair in comparing a Type I and a Type III graduation of certain material to use the same value of ν when there are four constants necessary to reach the former and three to reach the latter. He may ask 'are we not really restricting our freedom of choice more in the former case than the latter because, to take an extreme case, we should reproduce a distribution of only four groups exactly with Type I and alter it, that is have freedom of choice, if we use Type III?

8. Before we deal with this question we may explain that in the previous paragraphs of this chapter two distinct problems have been covered by the one word 'graduation'. These problems are:

I. Given a theoretical distribution, to ascertain the probability of getting an actual distribution or an equally likely or a less likely one.

On page 172 is an example which compares the theoretical number of 'heads', when six coins are tossed, with an actual distribution.

II. Given a graduation of an actual distribution, to ascertain the probability that the deviations will be the same as or greater than those found.

The answer depends on the number of constants in the formula used for graduation. If there are r constants we should deduct r from the number of groups instead of deducting unity as in the last example (see Section 4). Generally speaking, the same result is obtained if one more than the number of moments used in the calculations be deducted. This gives the theoretical answer to the question raised at the end of § 7 above. It is not always easy to interpret the number of moments in applying the rule: thus we choose between Type III and Type V by using the fourth moment, though there are only three moments needed to find the constants. Again if in Type I the start of the curve is fixed, three moments only are used; while if the range is fixed, only two moments are used (and in effect the number of constants is similarly decreased). If we make a rough attempt at a graduation by a Type I curve using four unadjusted moments and then vary the start of the curve as indicated in Chapter 5, § 11 of 'Additional Examples', then the final graduation only uses three moments. It can be argued that the full number of constants has been assumed and four moments have really been used. Certainly four *parameters* (in addition to N) have been estimated, since the start has, in fact, been found by trial and error.

No. of 'heads' (1)	Theoretical (2)	Actual (3)	(2)–(3) (4)	Square of (4) (5)	(5)/(2) (6)
0	1	0	1	1	1·00
1	6	6	0	0	0·00
2	15	12	3	9	0·60
3	20	23	−3	9	0·45
4	15	18	−3	9	0·60
5	6	3	3	9	1·50
6	1	2	−1	1	1·00
Total	64	64			$\chi^2 = 5{\cdot}15$

If $\nu = 6$ and $\chi^2 = 5{\cdot}15$, then $P = 0{\cdot}52$.

9. The example given for Problem I in § 8 can be used to explain the point mentioned in § 5 about undergraduation. We may, on a particular occasion, reach an actual distribution identical with the theoretical. χ^2 will then be zero and P will be unity. Similarly we may reach a distribution so far from the theoretical as to seem well-nigh impossible. One of these exceptional cases may appear and if we repeat the experiment long enough we shall get distributions giving all values of P. Similarly with graduation; we are unlikely, if we know the right form of curve, to find a value of P that is infinitesimally small or very near to unity, but neither is impossible.

10. When we merely want to compare several graduations of the same distribution we can often stop our work after the calculation of χ^2. Thus if we make graduations by Type I using various adjustments or compare them with Type A or Type B using the same number of constants, the lowest value of χ^2 shows the closest graduation. Even if the number of constants differs, the value of χ^2 shows which graduation is actually closest and for some actuarial work this may be more important than the study of the probabilities.

Bearing in mind that there are difficulties in interpreting the number of degrees of freedom in some cases, we may consider what is implied when we use the solution of Problem I for Problem II. All the old applications of the (P, χ^2) test were made in this way. In such circumstances we are saying, in effect, that the graduation is a theoretical distribution not necessarily obtained from the actual distribution but by general reasoning or from other previous experience, and that we are measuring the probability of divergences from that theoretical distribution as great as or greater than those of the actual distribution.

The points set out in these paragraphs are mentioned because it is well to be reminded that we must not read into a good general test of graduation a refinement which is neither justified by the underlying theory nor required in practical work.

11. Reference may here be made to a test of a graduation of a mortality table. The data are expressed as 'exposed to risk' (E_x) at each age (or group of ages) and 'deaths' (θ_x). A graduation of the rates of mortality is made and the 'expected deaths' (θ'_x) are calculated by multiplying the values of E_x by the appropriate graduated rates of mortality (q_x). We have, therefore, graduated the series

$$\theta_x,\ E_x-\theta_x;\ \theta_{x+1},\ E_{x+1}-\theta_{x+1};\ \text{etc.}$$

by

$$\theta'_x,\ E_x-\theta'_x;\ \theta'_{x+1},\ E_{x+1}-\theta'_{x+1};\ \text{etc.}$$

The E_x is fixed in each pair, so, if there are 40 ages, there are only 40 degrees of freedom, not 80. But the χ^2 should be calculated from all the 80 values, although when E is large relatively to θ, as it is at nearly every age, the $E-\theta$ terms give zero elements. It will be easier for the reader to follow this argument if he bears in mind that the total of the θ's need not be reproduced exactly by the θ''s. Deduction will have to be made from the 40 degrees of freedom for the number of constants used in the graduation.*

12. The χ^2 criterion does not use the information contained in the ordered pattern of the deviations (m'_r-m_r). Several test criteria which do make use of this information have been developed, and we will now briefly describe some of them. It will be well to bear in mind that the comments in § 5 also apply to these tests (where appropriate).

* A worked-out example is given on p. 237 of *Statistics: An Intermediate Text-Book*, Vol. 2 (N. L. Johnson and H. Tetley).

We will consider only the assessment of goodness-of-fit for univariate distributions. The most obvious feature of the pattern of deviations, which is ignored by the χ^2 test, is the sequence of *signs* of the deviations (since these are squared in the calculation of χ^2). A test can be applied to these signs alone. We count the number of *runs* of similar sign. Thus in the sequence

$$++----+--+-++++$$

there are 4 runs of $+$'s and 3 of $-$'s. Generally, in a sequence of $n_1 +$'s and $n_2 -$'s there can be 1, 2, ... or n_1 runs of $-$'s if $n_2 \geqslant n_1 - 1$. (If $n_2 < n_1 - 1$, the maximum possible number of runs of $+$'s is $n_2 + 1$.)

In order to find the chance of obtaining exactly g runs of $+$'s, we note that they can be regarded as being formed by inserting marks of partition at $(g-1)$ points of the possible $(n_1 - 1)$ positions between $+$'s in the sequence $+++....+++$ of n_1 $+$'s. These positions can be chosen in $\binom{n_1-1}{g-1}$ ways.* The n_2 $-$'s we have to split up into $(g+1)$ groups to be placed (in order) among the $+$'s at the points indicated by the marks of partition. The number of ways of doing this is $\binom{n_2-1}{g}$, but we can allow the first and/or last groups to be empty (i.e. contain no $-$). With the addition of these cases, the number of ways is $\binom{n_2+1}{g}$. The total number of arrangements, with exactly g groups of $+$'s out of $\binom{n_1+n_2}{n_1}$ possible arrangements, is $\binom{n_1-1}{g-1}\binom{n_2+1}{g}$. The chance of g groups of $+$'s is thus

$$\binom{n_1-1}{g-1}\binom{n_2+1}{g}\Big/\binom{n_1+n_2}{n_1}$$

Formally the distribution of g is that which would be obtained in the two-way table shown below (where n_1, n_2 are fixed) if the two classifications were independent. So it is natural to use the test for

* We use $\binom{N}{R}$ to denote the number of ways of choosing sets of R objects from N things. The value of $\binom{N}{R}$ is $N!/(R!(N-R)!)$, that is

$$[N(N-1)...(N-R+1)]/[R(R-1)...3\cdot 2]$$

n_1-g	$g-1$	n_1-1
g	n_2-g+1	n_2+1
n_1	n_2	n_1+n_2

independence appropriate to two-way tables. This is applied by regarding

$$(n_1+n_2)[(n_1+n_2)g-n_1(n_2+1)]^2[n_1n_2(n_1-1)(n_2+1)]^{-1}$$

as a χ^2 with one degree of freedom. This is just the square of a variable having a unit normal distribution (i.e. described by a unit normal frequency-curve).

This test and the χ^2 test can be combined in various ways.* To a good approximation they are independent. However, it is rather more informative to record separately the results of the two tests.

A test which takes into account the sequences of both signs and magnitudes of deviations is the 'smooth' test devised by J. Neyman.† We will describe the mode of application of the test, but will not go into much detail, as the test is not widely used.

First we suppose that the data are ungrouped. We then form the so-called 'probability integral transform'

$$y_r = \int_{-\infty}^{x_r} f_x\,dx$$

for each of the observed values x_r (where f_x is the curve for which goodness-of-fit is to be tested). We then calculate

$$\begin{aligned} u_1^2 &= 12T_1^2n^{-1} \\ u_2^2 &= 180(T_2-\tfrac{1}{12}n)^2n^{-1} \\ u_3^2 &= 7(20T_3-3T_1)^2n^{-1} \\ u_4^2 &= (210T_4-45T_2+9n/8)^2n^{-1} \end{aligned}$$

where $T_j = \sum(y_r-\frac{1}{2})^j$.

Then we calculate

P = probability that χ^2 with 4 degrees of freedom exceeds $(u_1^2+u_2^2+u_3^2+u_4^2)$.

This value of P is then used in the same way as the P calculated from the usual χ^2 test.

It is possible to use only the first three, or even two of the u^2's (with appropriate changes in the degrees of freedom of the χ^2).

* F. N. David, *Biometrika*, XXXIV (1947), 299.

† H. L. Seal, *Biometrika*, XXXV (1948), 202; D. A. S. Fraser, *Biometrika*, XXXVII (1950), 447.

Properties of this test have been studied by D. E. Barton.* In particular,† he studied its application to grouped data, and the effect of using estimates of parameters in the fitted curves.

13. Another test procedure is to compare the sample cumulative distribution function with that of the fitted distribution at all values of the available, and to take the maximum value of the absolute magnitude of the difference as a measure of disparity of these two functions. To a fair approximation ‡ the probability that this exceeds $zN^{-\frac{1}{2}}$ (N being the size of the sample) is

$$\alpha(z) = 2 \sum_{j=1}^{\infty} (-1)^{j-1} \exp(-2j^2z^2)$$

(if N is large).

Some values of z, for specified values of $\alpha(z)$ are

$\alpha(z)$	0·01	0·025	0·05	0·1
z	1·63	1·48	1·36	1·22

If the maximum observed difference exceeds $zN^{-\frac{1}{2}}$ (the value of z being chosen to correspond with a desired value of $\alpha(z)$), this may be taken as evidence of a lack of goodness of fit.

It is not easy to apply this test when the data are grouped, except on the basis of more or less plausible assumptions about distributions of data within groups. One can, of course, carry out a rough test by using an upper bound for the absolute difference, which can be determined easily from the data by assuming all observations in a group to be concentrated at one end of the group. If this upper bound itself is less than the critical value of $zN^{-\frac{1}{2}}$ we can be satisfied about goodness-of-fit, so far as this criterion is concerned.

* *Skand. Aktuar.* XXXVI (1953), 24; XXXIX (1956), 216. Also *Trabajos de Estadistica,* IV (1953), 199.

† *Skand. Aktuar.* XXXVIII (1955), 1–16.

‡ See N. Smirnov, *Ann. Math. Statist.* XIX (1948), 279. (Tables for smaller values of N have been given by F. J. Massey, *J. Amer. Statist. Assoc.*, XLVI (1951), 68.)

APPENDIX 1

CORRECTIONS FOR MOMENTS

1. The following method has been suggested by E. Pairman and K. Pearson (*Biometrika*, XII, 231 et seq.) when the curve rises abruptly at one or both ends.

Let n_1, n_2, etc. be the proportionate frequencies in the 1st, 2nd, etc. groups, then put

$$\begin{aligned}
a_1 &= -\tfrac{1}{60}\{137n_1-163n_2+137n_3-63n_4+12n_5\}\\
a_2 &= \tfrac{1}{12}\{45n_1-109n_2+105n_3-51n_4+10n_5\}\\
a_3 &= -\tfrac{1}{4}\{17n_1-54n_2+64n_3-34n_4+7n_5\}\\
a_4 &= \{3n_1-11n_2+15n_3-9n_4+2n_5\}\\
a_5 &= -\{n_1-4n_2+6n_3-4n_4+n_5\}
\end{aligned}$$

Similarly, values of b_1, b_2, etc. can be obtained from the other end of the distribution.

Then the values of the moments are as follows, where A is the distance of the start and B is the distance of the end of the distribution from the origin about which moments are calculated:

$$\begin{aligned}
\mu_1' &= \nu_1'+\{\tfrac{1}{12}(a_1-\tfrac{1}{60}a_3+\tfrac{1}{2520}a_5)+\tfrac{1}{12}(b_1-\tfrac{1}{60}b_3+\tfrac{1}{2520}b_5)\}\\
\mu_2' &= \nu_2'-\tfrac{1}{12}+\{-\tfrac{1}{120}(a_2-\tfrac{5}{126}a_4)+\tfrac{1}{6}A(a_1-\tfrac{1}{60}a_3+\tfrac{1}{2520}a_5)\}\\
\mu_3' &= \nu_3'-\tfrac{1}{4}\nu_1'+\{-\tfrac{1}{40}(a_1-\tfrac{5}{63}a_3+\tfrac{1}{240}a_5)\\
&\qquad -\tfrac{1}{40}A(a_2-\tfrac{5}{126}a_4)+\tfrac{1}{4}A^2(a_1-\tfrac{1}{60}a_3+\tfrac{1}{2520}a_5)\}\\
\mu_4' &= \nu_4'-\tfrac{1}{2}\nu_2'+\tfrac{7}{240}+\{\tfrac{1}{126}(a_2-\tfrac{7}{80}a_4)-\tfrac{1}{10}A(a_1-\tfrac{5}{63}a_3+\tfrac{1}{240}a_5)\\
&\qquad -\tfrac{1}{20}A^2(a_2-\tfrac{5}{126}a_4)+\tfrac{1}{3}A^3(a_1-\tfrac{1}{60}a_3+\tfrac{1}{2520}a^5)\}
\end{aligned}$$

and similar expressions in B and b's.

If the moments be taken about the start of the first group so that the first group is multiplied by powers of $\tfrac{1}{2}$, the second by powers of $\tfrac{3}{2}$ and so on, this expression is simplified so far as the a terms are concerned because the terms involving A vanish.

2. The method of reaching these adjustments starts with the Euler-Maclaurin expansion and assumes that the curve takes the form

$$1+\frac{a_1}{1!}(x-A)+\frac{a_2}{2!}(x-A)^2,\text{ etc.}$$

at the beginning and a similar form at the end. This leads to the values of the a's.

The differential coefficients at each end required in the Euler-Maclaurin expansion are then evolved and the result given is reached.

The frequency at the start is approximately

$$\frac{N}{60}\{137n_1-163n_2+137n_3-63n_4+12n_5\}$$

By means of this expression we can discover how nearly the frequency curve comes to zero at the ends of the range.

3. A few numerical examples may be given. The rule that the area, in the case of high contact, can be found by adding ordinates when tested by adding 12 ordinates of the normal curve calculated to 5 decimal places, gave 1·24998 instead of 1·25000. Nine ordinates of a Type III curve with high contact gave 24473 instead of 24475.

An example of the method of § 1 above is taken from the paper there cited. Moments for $\sqrt{x}\times 100\,000$ from $x = 0$ to $x = 10$ were calculated, the exact result being known. The proportional frequencies, which may be taken as the data, were:

$$\begin{array}{ll} n_1 = 0{\cdot}031623 & n_6 = 0{\cdot}111205 \\ n_2 = 0{\cdot}057820 & n_7 = 0{\cdot}120904 \\ n_3 = 0{\cdot}074874 & n_8 = 0{\cdot}129880 \\ n_4 = 0{\cdot}088665 & n_9 = 0{\cdot}138273 \\ n_5 = 0{\cdot}100571 & n_{10} = 0{\cdot}146185 \\ & \overline{\quad 1{\cdot}000000} \end{array}$$

From these figures

$$\begin{array}{ll} a_1 = -0{\cdot}0131\,0643 & b_1 = 0{\cdot}1499\,9857 \\ a_2 = -0{\cdot}0444\,8167 & b_2 = -0{\cdot}0074\,9283 \\ a_3 = 0{\cdot}0258\,4150 & b_3 = -0{\cdot}0003\,9450 \\ a_4 = -0{\cdot}0148\,8400 & b_4 = -0{\cdot}0000\,2600 \\ a_5 = 0{\cdot}0045\,0200 & b_5 = -0{\cdot}0000\,3800 \end{array}$$

$$a_1-\tfrac{1}{60}a_3+\tfrac{1}{2520}a_5 = -0{\cdot}0135\,3533$$

$$a_2-\tfrac{5}{126}a_4 = -0{\cdot}0438\,9104 \text{ and so on.}$$

Putting $A = 0$ and $B = 10$ and calculating moments about the start, we require for the a adjustments

$$\tfrac{1}{12}(a_1-\tfrac{1}{60}a_3+\tfrac{1}{2520}a_5) = -0{\cdot}0011\,2794$$

and the other adjustments in order are

$$-0{\cdot}0003\,6576 \qquad -0{\cdot}0003\,7846 \quad \text{and} \quad -0{\cdot}0003\,4269$$

For the b terms we have

$$\tfrac{1}{12}(b_1-\tfrac{1}{60}b_3+\tfrac{1}{2520}b_5) = 0{\cdot}0125\,0043$$

$$\tfrac{1}{6}B(b_1-\tfrac{1}{60}b_3+\tfrac{1}{2520}b_5) = 0{\cdot}2500\,0860$$

and the other terms in order give

3·7501 2825	50·0017 1000
−0·0000 6244	−0·0001 8731
−0·0374 6365	0·0037 5074
0·1500 2972	−0·0000 5945

Finally for the adjusted moments we reach

	Raw moments	With Sheppard's adjustment	With full adjustment	True value	
ν_1'	5·9880	5·9880	5·9994	6·0000	μ_1'
ν_2'	42·6900	42·6067	42·8570	42·8571	μ_2'
ν_3'	331·0854	329·5884	333·3349	333·3333	μ_3'
ν_4'	2698·7735	2677·4576	2727·2757	2727·2727	μ_4'

4. The method described above gives good results but is laborious. The approximations are less satisfactory in those cases where the first group does not relate to a complete unit base and the curve rises abruptly. The same authors gave a method for **J**-shaped distributions, but I should not use it as a simple approximation can be found by examining the exponential (Type X).

When statistics expressible by the exponential $y = y_0 e^{-x/\sigma}$ are stated in groups for each equal subrange h of x, the successive groups are $\int_0^h y_0 e^{-x/\sigma}\,dx$; $\int_h^{2h} y_0 e^{-x/\sigma}\,dx$; etc.; or $y_0\sigma(1-e^{-h/\sigma})$; $y_0\sigma(1-e^{-h/\sigma})$ $e^{-h/\sigma}$; $y_0\sigma(1-e^{-h/\sigma})\,e^{-2h/\sigma}$; etc.

These terms may also be regarded as a geometrical progression,* the first term being $y_0\sigma(1-e^{-h/\sigma})$ and the common ratio $e^{-h/\sigma}$. It follows that if we treat the areas as a geometrical progression extending to infinity, calculate the moments on this assumption and read the result as graduated terms of a geometrical progression, we shall reach correctly graduated areas, and we can subsequently write down the equation to the curve with little trouble.

Other points are however involved. Let us write the geometrical progression as ka^x and put $A = (1-a)^{-1}$, then the moments about its mean are

2nd moment A^2-A
3rd moment $2A^3-3A^2+A$
4th moment $9A^4-18A^3+10A^2-A$

* 'Geometrical progression' is used throughout to describe a discrete series and 'exponential curve' to describe a continuous one.

and if we work out β_1 and β_2, we get $4+h^2/\mu_2$ and $9+h^2/\mu_2$ respectively.

Using the exponential, the moments, etc. about the mean are: $\mu_2 = \sigma^2$, $\mu_3 = 2\sigma^3$, $\mu_4 = 9\sigma^4$, $\beta_1 = 4$, $\beta_2 = 9$.

Hence when we calculate moments, assuming that the statistics form a geometrical progression, whereas they are really areas from a curve, and seek to choose the type of curve from Pearson's criteria in his system, we shall reach a persistent error. For this purpose the β_1 and β_2 found from the statistics should be reduced by h^2/μ_2.

This rule can be used as an approximation in all J-shaped curves and will be found to give satisfactory results.

So far we have assumed that we know the start of the curve and that all the bases of the areas are of equal size. If this does not apply we can, in the case of an exponential curve, fit the curve, excluding the first (incomplete) term, and regard that term as related to an appropriate base extrapolated from the graduation of the remainder. This is an arbitrary arrangement but has practical advantages.

In other J-shaped curves in similar circumstances the first step would be to assume an exponential, to find therefrom approximately the base of the first incomplete group, and then assume that the area is concentrated at the middle point. This will generally give good results: the assumption of the exponential overstates the base and the assumption of half-way assumes a less rapidly falling curve than the J-shaped forms of Types I and III. There is therefore a balance of error.

Turning to the statistical side, the example on page 92 gives $\mu_2 = 2{\cdot}045$, $\beta_1 = 4{\cdot}629$, $\beta_2 = 9{\cdot}502$. These figures come from the unadjusted moments, and deducting 0·49 from the above values for β_1 and β_2 we reach 4·14 and 9·01. The theoretical values when an exponential curve is to be used are 4 and 9.

If we apply the rule as an approximation in other J-shaped cases we find that in the example on page 95, where a twisted J-shaped curve is given, $\mu_2 = 4{\cdot}266$, $\beta_1 = 0{\cdot}761$, $\beta_2 = 2{\cdot}646$, and the adjustment leads to $\beta_1 = 0{\cdot}527$ and $\beta_2 = 2{\cdot}412$. Hence $5\beta_2 - 6\beta_1 - 9$ becomes $-0{\cdot}098$ instead of $-0{\cdot}368$. The theoretical criterion would lead us to expect $5\beta_2 - 6\beta_1 - 9 = 0$.

These examples are not, of course, complete evidence, but they show that the suggestion may lead to accurate results, and it has the merit of simplicity. The rule with regard to the adjustment of the

β's by h^2/μ_2 may be combined with the approximations given on page 92, where it is mentioned that the mean is overstated, when μ_3 is positive, by about $h^2/12\sigma$, and the second moment about the *true* mean (i.e. the mean as corrected by $h^2/(12\sigma)$) is understated by about $h^2/12$. We do not know σ exactly but can use the square root of the second moment as found from the calculations. If h be taken as a unit and the moments found in terms of h, i.e. in working units, the corrections are $1/(12\sqrt{\mu_2})$ and $\frac{1}{12}$.

5. An alternative to the method of § 1 is to find mid-ordinates corresponding to the areas of the groups and treat these mid-ordinates in the manner explained in Chapter 3, § 18.

The mid-ordinates m_1, m_2, etc. are found by the following equations:

$$m_3 = \tfrac{1}{1920}\{2134n_3 - 116(n_2+n_4) + 9(n_1+n_5)\}$$
$$m_2 = \tfrac{1}{1920}\{-71n_1 + 2044n_2 - 26n_3 - 36n_4 + 9n_5\}$$
$$m_1 = \tfrac{1}{1920}\{1689n_1 + 684n_2 - 746n_3 + 364n_4 - 71n_5\}$$

The total frequency is not exactly reproduced but the moments obtained are good approximations.

6. It has been pointed out that one of the difficulties in calculating moments when the curve rises abruptly at one or both ends arises because the true start or end of the curve is unknown. In other words, the base of the first area or last area (or both) is smaller than that of the other areas. In practice good results can often be obtained with unadjusted moments but the first attempt may require modification by varying the range of the curve (see p. 107). When this is done, the moment contribution for the first area, or the last, or both, must be recalculated by assuming that the area is concentrated at the middle of the smaller base.

E. S. Martin has approached the problem more systematically in a paper in *Biometrika*, XXIV, 12, and has given tables from which the start of the curve may be estimated.

APPENDIX 2

B AND Γ FUNCTIONS

$$\mathrm{B}(m,n) = \int_0^1 x^{m-1}(1-x)^{n-1}\,dx$$

$$\Gamma(p) = \int_0^\infty e^{-x}\,x^{p-1}\,dx$$

I. $$\int x^{p-1}\,e^{-x}\,dx = -e^{-x}\,x^{p-1}+(p-1)\int x^{p-2}\,e^{-x}\,dx$$

by integration by parts.

When $p-1$ is positive, $e^{-x}x^{p-1}$ vanishes when $x = 0$, and when $x = \infty$ it can be written x^{p-1}/e^x, and the rule for evaluation of undetermined forms can be applied giving zero as the result.

$$\int_0^\infty x^{p-1}\,e^{-x}\,dx = (p-1)\int_0^\infty x^{p-2}\,e^{-x}\,dx$$

or $$\Gamma(p) = (p-1)\Gamma(p-1)$$

If p be an integer, $\Gamma(p) = (p-1)!$

II. To prove $\mathrm{B}(m,n) = \dfrac{\Gamma(m)\Gamma(n)}{\Gamma(m+n)}$.

Putting zx for x in the equation for $\Gamma(m)$, we have

$$\Gamma(m) = \int_0^\infty e^{-zx}\,z^m\,x^{m-1}\,dx$$

and $$\Gamma(m)\,e^{-z}\,z^{n-1} = \int_0^\infty e^{-z(1+x)}\,z^{m+n-1}x^{m-1}\,dx$$

$$\therefore\quad \Gamma(m)\int_0^\infty e^{-z}\,z^{n-1}\,dz = \int_0^\infty\left[\int_0^\infty e^{-z(1+x)}\,z^{m+n-1}\,dz\right]x^{m-1}\,dx$$

But if $z(1+x) = y$, we get

$$\int_0^\infty e^{-z(1+x)}\,z^{m+n-1}\,dz = \frac{1}{(1+x)^{m+n}}\int_0^\infty e^{-y}\,y^{m+n-1}\,dy$$

$$= \frac{\Gamma(m+n)}{(1+x)^{m+n}}$$

$$\therefore\quad \Gamma(m)\Gamma(n) = \Gamma(m+n)\int_0^\infty \frac{x^{m-1}}{(1+x)^{m+n}}\,dx$$

But putting $1+x = 1/(1-z)$ in this integral, we obtain

$$\int_0^1 \frac{z^{m-1}}{(1-z)^{m-1}}(1-z)^{m+n}\,\frac{1}{(1-z)^2}\,dz$$

which reduces at once to $B(m, n)$, and therefore

$$B(m,n) = \frac{\Gamma(m)\Gamma(n)}{\Gamma(m+n)}$$

III. To prove $\Gamma(\frac{1}{2}) = \sqrt{\pi}$.

We have already shown (see p. 74) that $2\int_0^\infty e^{-x^2}\,dx = \sqrt{\pi}$, and by putting $x^2 = z$, we have

$$2\int_0^\infty e^{-x^2}\,dx = \int_0^\infty e^{-z}\,z^{-\frac{1}{2}}\,dz = \Gamma(\tfrac{1}{2}) = \sqrt{\pi}$$

For statistical work a table of $\Gamma(x)$ or $\log \Gamma(x)$ is required (see pp. 210–1 or, better still, J. Brownlee, 'Log $\Gamma(x)$ from $x = 1$ to 50·9 by intervals of 0·01', (*Tracts for Computers*, No. IX, Camb. Univ. Press), or pp. 267–70 of *Handbook of Mathematical Tables* (Ed. Abramowitz and Stegun, Nat. Bur. Standards Applied Math. Series No. 55—U.S. Govt. Printing Office) where values of $\Gamma(x)$ and $\log \Gamma(x)$ are given to ten decimal places for $x = 1{\cdot}000\ (0{\cdot}005)\ 2{\cdot}000$).

When x is large, we can approximate to $\log \Gamma(x)$. The best known approximation is

$$\Gamma(x+1) = \sqrt{(2\pi x)}x^x\,e^{-x}e^{\,1/12x}*$$

or

$$\log_{10}\Gamma(x+1) = \log_{10}\sqrt{(2\pi)}+(x+\tfrac{1}{2})\log_{10}x-\left(x-\frac{1}{12x}\right)\log_{10}e$$

and it can be used when x is not less than 8. The calculations on page 184 show how the table of $\log \Gamma(x)$ is used, and how the approximation approaches the true value.

The functions already described may be described as 'complete Γ-functions' and 'complete B-functions' but for certain purposes 'incomplete' functions are required. Thus, if we have to express the areas of a Type III curve we require quantities like $\int_a^b e^{-x}\,x^p\,dx$, and these integrals can be obtained by subtracting $\int_0^a e^{-x}\,x^p\,dx$ from $\int_0^b e^{-x}\,x^p\,dx$. The incomplete Γ function, $\Gamma_x(p+1)$, is defined as $\int_0^x e^{-x}\,x^p\,dx$. Bearing in mind that y_0 involves the corresponding

* A proof of this well-known approximation will be found in *Chrystal's Algebra*, II, 368, etc., or in *Boole's Finite Differences*, Chapter VI.

complete Γ function in the denominator and the values assumed by Γ functions, it has been found preferable to tabulate the ratio $\Gamma_x(p+1)/\Gamma(p+1)$. This function of x and p always lies between 0 and 1, and expresses the chance of a value less than a given limit if

x	Log $\Gamma(x)$		
	True	Approximate	Δ
1·372	$\bar{1}$·948975	...	...
2·372	0·086329	0·086743	0·000414
3·372	0·461444	0·461532	0·000088
4·372	0·989332	0·989369	0·000037
5·372	1·630012	1·630025	0·000013
6·372	2·360148	2·360156	0·000008
7·372	3·164424	3·164430	0·000006
8·372	4·032009	4·032010	0·000001
9·372	4·954838	4·954837	−0·000001
10·372	5·926670	5·926669	−0·000001

the frequency is expressible by a Type III curve. There is however a difficulty in tabulation which has been met by using, instead of x, the argument $u = x/\sqrt{(p+1)}$. Tables of

$$I(u,p) = \int_0^{u\sqrt{p+1}} e^{-x}\, x^p\, dx/\Gamma(p+1)$$

have been published.*

The incomplete B function is defined as

$$\mathrm{B}_x(p,q) = \int_0^x x^{p-1}(1-x)^{q-1}\,dx$$

and it is tabulated and published † in the form

$$I_x(p,q) = \mathrm{B}_x(p,q)/\mathrm{B}(p,q)$$

Though the tables of the incomplete Γ and B functions are valuable for certain purposes, the student will find that for ordinary curve-fitting the methods described in Chapter 5 will suffice.

* *Tables of the Incomplete Γ-Function.* Printed by Cambridge University Press and published by *Biometrika.*

† *Tables of the Incomplete Beta-Function.* Published by *Biometrika.*

APPENDIX 3

THE EQUATION TO THE NORMAL SURFACE

Let $\eta_1, \eta_2, \eta_3, \ldots \eta_n$ be deviations from their respective means of a complex of measurable characteristics. The sizes of the functions measured, or organs, are determined by a large number of *independent* contributory causes. Let there be m of these causes, and let their deviations from their means be $\epsilon_1, \epsilon_2, \epsilon_3, \ldots \epsilon_m$, then $\eta_1, \eta_2, \eta_3, \ldots \eta_n$ will be functions of $\epsilon_1, \epsilon_2, \epsilon_3, \ldots \epsilon_m$. Further, if $m > n$ the ϵ's will not be fully determined for a given η complex. We also assume that the variations in intensity of the contributory causes are small as compared with their absolute intensity, and that these variations follow the normal law of distribution; that is, we assume that the deviations from the mean value can be graduated by the normal curve of error. The mean complex being reached with the mean intensities of contributory causes, we have, by the principle of the super position of small quantities,

$$\left.\begin{aligned} \eta_1 &= a_{11}\epsilon_1 + a_{12}\epsilon_2 + a_{13}\epsilon_3 + \ldots + a_{1m}\epsilon_m \\ \eta_2 &= a_{21}\epsilon_1 + a_{22}\epsilon_2 + a_{23}\epsilon_3 + \ldots + a_{2m}\epsilon_m \\ &\ldots\ldots\ldots\ldots\ldots\ldots\ldots\ldots\ldots\ldots \\ \eta_n &= a_{n1}\epsilon_1 + a_{n2}\epsilon_2 + a_{n3}\epsilon_3 + \ldots + a_{nm}\epsilon_m \end{aligned}\right\} \qquad \ldots\ldots(1)$$

The a's are coefficients whose values have to be determined, and any of the system of a's may be zero, for a particular contributory cause may have no effect on a particular result. Further, the chance that we have a conjunction of contributory causes lying between ϵ_1 and $\epsilon_1 + \delta\epsilon_1$, ϵ_2 and $\epsilon_2 + \delta\epsilon_2$, ... and between ϵ_m and $\epsilon_m + \delta\epsilon_m$ will be given approximately by

$$P' = K e^{-\left(\frac{\epsilon_1^2}{2\kappa_1^2} + \frac{\epsilon_2^2}{2\kappa_2^2} + \ldots + \frac{\epsilon^2{}_m}{2\kappa^2{}_m}\right)} \times \delta\epsilon_1 \delta\epsilon_2 \ldots \delta\epsilon_m \qquad \ldots\ldots(2)$$

where the standard deviations of the distributions are $\kappa_1, \kappa_2, \ldots \kappa_m$ and K is constant.*

Now, by (1) let n of the variables ϵ, say the first n, be replaced by the variables η, then the probability that we have a complex with

* Considering any particular case of the normal curve, the chance of getting a result between $\epsilon_1 + \delta\epsilon_1$ when the distribution is 'normal' is approximately $y_0 e^{-\epsilon_1^2/2\kappa_1^2} \delta\epsilon_1$ where κ_1 is the standard deviation; similarly with each of the other causes. As the causes are independent, the product of the various chances gives the required chance.

organs lying between η_1 and $\eta_1+\delta\eta_1$, η_2 and $\eta_2+\delta\eta_2$, ... η_n and $\eta_n+\delta\eta_n$, together with a series of contributory causes lying between ϵ_{n+1} and $\epsilon_{n+1}+\delta\epsilon_{n+1}$, ϵ_{n+2} and $\epsilon_{n+2}+\delta\epsilon_{n+2}$, ... ϵ_m and $\epsilon_m+\delta\epsilon_m$, will be

$$P' = C' e^{-\frac{1}{2}\phi^2}\delta\eta_1\delta\eta_2\ldots\delta\eta_n\delta\epsilon_{n+1}\delta\epsilon_{n+2}\ldots\delta\epsilon_m$$

where C' is a constant, a function of K and the a's, and ϕ^2 consists of the following parts:

(i) A quadratic function of the η's from η_1 to η_n.

(ii) A quadratic function of the ϵ's from ϵ_{n+1} to ϵ_m.

(iii) A series of functions of the type

$$\epsilon_{n+1}(b_{1,n+1}\eta_1+b_{2,n+1}\eta_2+\ldots+b_{n,n+1}\eta_n)$$
$$\epsilon_{n+2}(b_{1,n+2}\eta_1+b_{2,n+2}\eta_2+\ldots+b_{n,n+2}\eta_n)$$
$$\ldots\ldots\ldots\ldots\ldots\ldots\ldots\ldots\ldots\ldots\ldots\ldots\ldots\ldots\ldots$$
$$\epsilon_m(b_{1,m}\eta_1+b_{2,m}\eta_2+\ldots+b_{n,m}\eta_n)$$

where some of the b's may be zero.

Now, if P' be integrated for all values from $-\infty$ to $+\infty$ of all the contributory causes ϵ_{n+1}, ϵ_{n+2}, ... ϵ_m, we shall have the whole chance of a complex with organs falling between η_1 and $\eta_1+\delta\eta_1$, η_2 and $\eta_2+\delta\eta_2$, ... η_n and $\eta_n+\delta\eta_n$. But every time we integrate with regard to an ϵ, say ϵ_{n+1}, we alter the constants of each contributory part of ϕ^2, but do not alter the triple constitution of ϕ^2 except to cause one ϵ to disappear from its (ii) and (iii) constituents. At the same time we alter C' without introducing into it any terms in η. Thus, finally, after $m-n$ integrations, ϕ^2 is reduced to its first constituent, or we conclude that the chance of a complex of organs between η_1 and $\eta_1+\delta\eta_1$, η_2 and $\eta_2+\delta\eta_2$, ... η_n and $\eta_n+\delta\eta_n$ occurring is given by

$$P = C e^{-\frac{1}{2}\chi^2}\delta\eta_1\delta\eta_2\delta\eta_3\ldots\delta\eta_n \qquad \ldots\ldots(3)$$

where χ^2 is a quadratic function of the η's. This is the law of frequency for the complex.

Consider the expression (3) but replace χ^2 by a quadratic function, then

$$P = C e^{-\frac{1}{2}\{c_{11}\eta_1{}^2+c_{22}\eta_2{}^2+\ldots+2c_{12}\eta_1\eta_2+2c_{13}\eta_1\eta_3+\ldots\}}$$
$$= C e^{-\frac{1}{2}\{S_1(c_{pp}\eta_p{}^2)+2S_2c_{pq}\eta_p\eta_q)\}}$$

Here C, c_{pp}, c_{pq} are constants, and S_1 denotes a summation for every value of p, and S_2 for every pair of values of p and q in the series.

Taking the simplest case of two variables, this becomes, with slight modifications of notation,

$$P = C e^{-\frac{1}{2}(c_1\eta_1{}^2+c_2\eta_2{}^2+2c_{12}\eta_1\eta_2)}$$

We now assume that neither c_1 nor c_2 is zero.

Integrate P for all values of η_1 from $-\infty$ to $+\infty$, and we must have the normal curve of η_2 variation.* Therefore

$$\frac{1}{2\sigma_1^2} = c_2\left(1 - \frac{c_{12}^2}{c_1 c_2}\right)$$

Similarly, integrating for all values of η_2,

$$\frac{1}{2\sigma_2^2} = c_1\left(1 - \frac{c_{12}^2}{c_1 c_2}\right)$$

Integrating for all values of η_1 and η_2 to obtain the total frequency, we have

$$N = \frac{2C\pi}{\sqrt{(c_1 c_2 - c_{12}^2)}}$$

Now put $\rho = -c_{12}/\sqrt{(c_1 c_2)}$ and write x and y for η_1 and η_2, and we have

$$Z = \frac{N}{2\pi\sigma_1\sigma_2\sqrt{(1-\rho^2)}} e^{-\frac{1}{2}\left\{\frac{x^2}{\sigma_1^2(1-\rho^2)} - \frac{2xy\rho}{\sigma_1\sigma_2(1-\rho^2)} + \frac{y^2}{\sigma_2^2(1-\rho^2)}\right\}} \qquad \ldots\ldots(4)$$

* The result $1/2\sigma_1^2 = c_2(1 - c_{12}^2/c_1c_2)$ can be reached at once by re-arranging the index in the expression of P as {a perfect square $-\frac{1}{2}c_2(1-c_{12}^2/c_1c_2)\eta_2^2$}.

APPENDIX 4

THE INTEGRATION OF SOME EXPRESSIONS CONNECTED WITH THE NORMAL CURVE OF ERROR

1. On page 74 we showed that

$$\int_{-\infty}^{\infty} e^{-x^2}\,dx = \sqrt{\pi} \qquad \text{......(1)}$$

$$\therefore \qquad \int_{-\infty}^{\infty} e^{-x^2/h^2}\,dx = h\sqrt{\pi} \qquad \text{......(2)}$$

Since the curve is symmetrical, we have

$$\int_{-\infty}^{+\infty} x^{2n+1}\, e^{-x^2}\,dx = 0 \qquad \text{......(3)}$$

If we integrate $\int x^{2n}\, e^{-x^2}\,dx$ by parts, we have

$$\int x^{2n}\, e^{-x^2}\,dx = \frac{x^{2n+1}}{2n+1}\, e^{-x^2} + \int 2x \frac{x^{2n+1}}{2n+1}\, e^{-x^2}\,dx$$

and inserting the limits $-\infty$ and $+\infty$

$$\int_{-\infty}^{+\infty} x^{2n}\, e^{-x^2}\,dx = \frac{2}{2n+1}\int_{-\infty}^{+\infty} x^{2n+2}\, e^{-x^2}\,dx \qquad \text{......(4)}$$

This last formula shows the connection between the successive even moments.

2. Let

$$Z = Z_0\, e^{-\frac{1}{2(1-\rho^2)}\left\{\frac{x^2}{\sigma_1^2} - \frac{2xy\rho}{\sigma_1\sigma_2} + \frac{y^2}{\sigma_2^2}\right\}}$$

Then Z can be put in the form

$$Z_0 e^{-\frac{1}{2(1-\rho^2)}\left\{\frac{x}{\sigma_1} - \frac{\rho y}{\sigma^2}\right\}^2} e^{-\frac{1}{2(1-\rho^2)}\cdot\frac{y^2(1-\rho^2)}{\sigma_2^2}} = Z_0 e^{-\frac{1}{2(1-\rho^2)\sigma_1^2}\left\{x - \frac{y\rho\sigma_1}{\sigma_2}\right\}^2} e^{-\frac{y^2}{2\sigma_2^2}}$$

Then $\displaystyle\int_{-\infty}^{+\infty} Z\,dx = Z_0\sqrt{\{2\pi(1-\rho^2)\}}\,\sigma_1\, e^{-y^2/2\sigma_2^2}$ by (2)

and

$$N = \int_{-\infty}^{+\infty}\int_{-\infty}^{+\infty} Z\,dx\,dy = \int_{-\infty}^{+\infty} Z_0\sqrt{\{2\pi(1-\rho^2)\}}\,\sigma_1\, e^{-y^2/2\sigma_2^2}\,dy$$

$$= 2\pi\sigma_1\sigma_2\sqrt{(1-\rho^2)}\,Z_0 \qquad \text{......(5)}$$

3. Using the same method as that just given, it can be shown that if $ac>b^2$

$$\int_{-\infty}^{+\infty}\int_{-\infty}^{+\infty} e^{-\frac{1}{2}(ax^2-2bxy+cy^2)}\,dx\,dy = \frac{2\pi}{\sqrt{(ac-b^2)}} \qquad \text{......(6)}$$

for the index can be written

$$-\frac{a}{2}\left\{x-\frac{b}{a}y\right\}^2+\frac{y^2}{2a}(ac-b^2)$$

and if we then integrate with respect to x, we have

$$\sqrt{\frac{2\pi}{a}}\,e^{-\frac{y^2}{2a}(ac-b^2)}$$

and if $ac-b^2$ is positive, we can integrate this last expression with respect to y and have

$$\sqrt{\frac{2\pi}{a}}\sqrt{\frac{2\pi a}{ac-b^2}}=\frac{2\pi}{\sqrt{(ac-b^2)}}$$

4. We shall now find $\displaystyle\int_{-\infty}^{+\infty}\int_{-\infty}^{+\infty} Zxy\,dx\,dy.$

Proceeding as in (5), we have

$$\int_{-\infty}^{+\infty} Zxy\,dx = \int_{-\infty}^{+\infty} xyZ_0 e^{-y^2/2\sigma_2^2}\,e^{-\frac{1}{2(1-\rho^2)\sigma_1^2}\left\{x-\frac{y\rho\sigma_1}{\sigma_2}\right\}^2}\,dx$$

$$= Z_0 y\,e^{-y^2/2\sigma_2^2}\int_{-\infty}^{+\infty}\left(X+\frac{y\rho\sigma_1}{\sigma_2}\right)e^{-\frac{X^2}{2(1-\rho^2)\sigma_1^2}}\,dX$$

where $X = x-\dfrac{y\rho\sigma_1}{\sigma_2}$

$$= Z_0 y\,e^{-y^2/2\sigma_2^2}\int_{-\infty}^{+\infty}\frac{y\rho\sigma_1}{\sigma_2}\,e^{-\frac{X^2}{2(1-\rho^2)\sigma_1^2}}\,dX$$

because by (3) $\displaystyle\int_{-\infty}^{+\infty} e^{-X^2/h^2}X\,dX$ is zero

$$= Z_0\frac{\sigma_1^2}{\sigma_2}\rho\sqrt{\{2\pi(1-\rho^2)\}}\,y^2 e^{-y^2/2\sigma_2^2}$$

But, by putting $n = 0$ in (4) and using (2), we have

$$\int_{-\infty}^{+\infty} y^2 e^{-y^2/2\sigma_2^2}\,dy = \sigma_2^3\sqrt{(2\pi)}$$

$$\therefore\quad \int_{-\infty}^{+\infty}\int_{-\infty}^{+\infty} Zxy\,dx\,dy = Z_0\sigma_1^2\sigma_2^2 2\pi\rho\sqrt{(1-\rho^2)}$$

$$= N\sigma_1\sigma_2\rho \qquad \text{......(7)}$$

because by (5)

$$Z_0 = \frac{N}{2\pi\sigma_1\sigma_2\sqrt{(1-\rho^2)}}$$

5. We may now deal with the problem referred to in Chapter 9. 'To find a value for r from the equation

$$\frac{N}{2\pi\sqrt{(1-r^2)}}\int_k^\infty\int_h^\infty e^{-\frac{1}{2}\frac{1}{1-r^2}(x^2+y^2-2rxy)}\,dx\,dy = d$$

where d, N, h, and k are known.'

Consider the expression

$$\frac{1}{\sqrt{(1-r^2)}}e^{-(x^2+y^2-2rxy)/2(1-r^2)} = U, \text{ say} \qquad \text{......}(\alpha)$$

and expand it in terms of r by Maclaurin's theorem, then

$$U = e^{-\frac{1}{2}(x^2+y^2)}\left(u_0+\frac{u_1r}{1!}+\frac{u_2r^2}{2!}+\ldots+\frac{u_nr^n}{n!}+\ldots\right) \qquad \text{......}(\beta)$$

where $$u_n = e^{\frac{1}{2}(x^2+y^2)}\left(\frac{d^nU}{dr^n}\right)_{r=0} \qquad \text{......}(\gamma)$$

Now take the logarithmic differential coefficient of U with respect to r, and we have

$$\frac{1}{U}\frac{dU}{dr} = -\frac{x^2+y^2-2rxy}{2}\frac{d}{dr}(1-r^2)^{-1}-$$
$$\frac{(1-r^2)^{-1}}{2}\frac{d}{dr}(x^2+y^2-2rxy)-\frac{1}{2}\frac{d}{dr}\log(1-r^2)$$
$$= -(x^2+y^2-2rxy)r(1-r^2)^{-2}+(1-r^2)^{-1}xy+r(1-r^2)^{-1}$$

$$\therefore \qquad (1-r^2)^2\frac{dU}{dr} = U\{xy+r(1-x^2-y^2)+r^2xy-r^3\}$$

Differentiate n times by Leibnitz's theorem and put $r = 0$ and we have

$$u_{n+1} = n(2n-1-x^2-y^2)u_{n-1}-n(n-1)(n-2)^2u_{n-3}$$
$$+xy\{u_n+n(n-1)u_{n-2}\}$$

Hence

$$\left.\begin{aligned} u_0 &= 1\\ u_1 &= xy\\ u_2 &= (x^2-1)(y^2-1)\\ u_3 &= x(x^2-3)(y^2-3)y\\ u_4 &= (x^4-6x^2+3)(y^4-6y^2+3)\end{aligned}\right\} \qquad \text{......}(\delta)$$

The laws indicated by (δ) are

$$\left.\begin{aligned} u_n &= v_n \times w_n \\ v_n &= xv_{n-1}-(n-1)v_{n-2} \\ w_n &= yw_{n-1}-(n-1)w_{n-2} \end{aligned}\right\} \qquad \ldots\ldots(\epsilon)$$

and we can, therefore, re-write (β) as

$$\frac{1}{2\pi}U = \frac{1}{2\pi}e^{-\frac{1}{2}(x^2+y^2)}\left(1+\frac{v_1 w_1}{1!}r+\frac{v_2 w_2}{2!}r^2+\ldots\right) \qquad \ldots\ldots(\zeta)$$

Integrating this from h to ∞ with respect to x, and remembering that w_n does not involve x, we have

$$\frac{1}{2\pi}\int_h^\infty U\,dx = \frac{1}{2\pi}e^{-\frac{1}{2}y^2}\left\{\int_h^\infty e^{-\frac{1}{2}x^2}\,dx+\frac{rw_1}{1!}\int_h^\infty e^{-\frac{1}{2}x^2}v_1\,dx+\ldots\right.$$

$$\left.+\frac{r^n w_n}{n!}\int_h^\infty e^{-\frac{1}{2}x^2}v_n\,dx+\ldots\right\}$$

$$= \frac{1}{\sqrt{(2\pi)}}e^{-\frac{1}{2}y^2}\left\{V_0+\frac{rV_1 w_1}{1!}+\ldots+\frac{r^n V_n w_n}{n!}+\ldots\right\}$$

where V_n is written for $\dfrac{1}{\sqrt{(2\pi)}}\displaystyle\int_h^\infty v_n e^{-\frac{1}{2}y^2}\,dx.$

Now integrate this with respect to y from k to ∞, remembering that V_n does not involve y, and writing W_n for $\dfrac{1}{\sqrt{(2\pi)}}\displaystyle\int_k^\infty e^{-\frac{1}{2}y^2}w_n\,dy$, we see that $\dfrac{1}{2\pi}\displaystyle\int_k^\infty\int_h^\infty U\,dx\,dy$ can be expressed as a series of which the general term is $(r^n/n!)V_n W_n$, and we must now evaluate V_n and W_n.

From (ϵ) it can be shown by induction that the general form of v_n is

$$x^n-n_2x^{n-2}+3n_4x^{n-4}-5n_6x^{n-6}+\ldots \qquad \ldots\ldots(\eta)$$

where $\qquad n_t = n(n-1)(n-2)\ldots(n-t+1)/t!$

Now we notice that

$$\frac{dv_n}{dx} = nv_{n-1} \qquad \ldots\ldots(\theta)$$

$\therefore$ by (ϵ) $\qquad v_n = xv_{n-1} - \dfrac{dv_{n-1}}{dx}$

Multiply by $e^{-\frac{1}{2}x^2}$ and integrate

$$\int e^{-\frac{1}{2}x^2}v_n\,dx = \int x\,e^{-\frac{1}{2}x^2}v_{n-1}\,dx-\int e^{-\frac{1}{2}x^2}\frac{dv_{n-1}}{dx}\,dx$$

and integrating the latter integral by parts, we have

$$\int e^{-\frac{1}{2}x^2} v_n \, dx = -e^{-\frac{1}{2}x^2} v_{n-1}$$

or
$$V_n = \frac{1}{\sqrt{(2\pi)}} \int_h^\infty v_n \, e^{-\frac{1}{2}x^2} dx = \frac{1}{\sqrt{(2\pi)}} e^{-\frac{1}{2}h^2} (v_{n-1})_{x=h}$$

Now, writing H for $\frac{1}{\sqrt{(2\pi)}} e^{-\frac{1}{2}h^2}$, and K for $\frac{1}{\sqrt{(2\pi)}} e^{-\frac{1}{2}k^2}$, we have from ($\alpha$)

$$\begin{aligned}
\frac{d}{N} &= \frac{1}{2\pi} \int_k^\infty \int_h^\infty U \, dx \, dy \\
&= \frac{1}{2\pi} \int_h^\infty \int_k^\infty e^{-\frac{1}{2}(x^2+y^2)} dx \, dy + \overset{\infty}{\underset{1}{S}} \left(\frac{r^n}{n!} HK (v_{n-1})_{x=h} (w_{n-1})_{y=k} \right) \\
&= \frac{(b+d)(c+d)}{N^2} + \overset{\infty}{\underset{1}{S}} \left(\frac{r^n}{n!} HK (v_{n-1})_{x=h} (w_{n-1})_{y=k} \right)
\end{aligned}$$

or remembering that $N = a+b+c+d$, we write

$$\begin{aligned}
\frac{ad-bc}{N^2 HK} &= \overset{\infty}{\underset{1}{S}} \left(\frac{r^n}{n!} (v_{n-1})_{x=h} \times (w_{n-1})_{y=k} \right) \\
&= r + \frac{r^2}{2} hk + \frac{r^3}{6} (h^2-1)(k^2-1) + \frac{r^4}{24} h(h^2-3)k(k^2-3) \\
&\quad + \frac{r^5}{120} (h^4-6h^2+3)(k^4-6k^2+3) \\
&\quad + \frac{r^6}{720} h(h^4-10h^2+15)k(k^4-10k^2+15) \\
&\quad + \frac{r^7}{5040} (h^6-15h^4+45h^2-15)(k^6-15k^4+45k^2-15) \\
&\quad + \text{etc.} \ldots \qquad \ldots\ldots(8)
\end{aligned}$$

APPENDIX 5

OTHER METHODS OF FITTING CURVES

The method of moments described in Chapter 3 is not the only method available for fitting curves. As would be expected of any general method, the method of moments may give difficulty because (i) the equations resulting from

$$\int_h^k x^n f(x, a, b, \ldots)\, dx = \nu_n$$

cannot be solved to find the constants $a, b, \ldots$ in $y = f(x, a, b, \ldots)$ or (ii) sufficiently accurate adjustments are not yet available in all circumstances for the statistical moments or (iii) some of the moments are liable to such large standard errors as to make them unreliable. This last point can be exemplified by curves, within the Pearson system, which cannot be fitted satisfactorily by moments, e.g. $y = y_0(1+x^2)^{-1}$.

The method of least squares

This method of adjusting observations makes the sum of the squares of the differences between the actual facts and the adjusted figures a minimum. The theory underlying the method is that errors are distributed in accordance with the 'normal curve of error'.

An example of the application of the method to fitting

$$a+bx+cx^2+\ldots$$

where the observations are of equal weight would make

$$S\{y-(a+bx+cx^2+\ldots)\}^2$$

a minimum where y stands for the observation. The equations for finding a, b, c, etc. would be

$$\begin{aligned} S(y) &= aS(1)+bS(x)+cS(x^2)+\ldots \\ S(xy) &= aS(x)+bS(x^2)+cS(x^3)+\ldots \\ S(x^2y) &= aS(x^2)+bS(x^3)+cS(x^4)+\ldots \end{aligned}$$

and so until we have sufficient equations: then these equations have to be solved for a, b, c, etc.

The method has been extended to cases where a definite condition holds (e.g. the three angles of a triangle are measured and have to be adjusted subject to the condition that the total is equal to two right angles) or where observations have unequal weight. The reader who wishes to pursue this line of work may be referred to one of the textbooks on the subject.*

The weakness of this method for fitting curves is that it often leads to equations which are incapable of solution. This can be seen by writing and trying to solve equations similar to those set out above when $f(x)$ is taken as, say, a frequency curve of Type III instead of $a+bx+cx^2+\dots$. It follows that for general curve fitting the method of moments is preferable.† When the two methods can be compared conveniently they both give good results—probably, equally good. It is sometimes asserted that the method of least squares must give the best result. If the test of 'best' is the smallest sum of the squares of the differences, then, obviously, the result is the best, but this is arguing in a circle. The difficulty lies in defining 'best'. A fair summary as regards curve fitting is that both methods give very good results but the method of moments is more generally applicable.

The method of maximum likelihood

This is due to R. A. Fisher.‡ If a distribution is to be represented by $y = f(x, a, b, \dots)$, where $a, b, \dots$ are the unknown constants and the chance of an observation falling in the range dx is represented by $f(x, a, b, \dots)\,dx$, then the chance that in a sample of n there will be n_1 in the range dx_1, n_2 in the range dx_2, and so on, is

$$\frac{n!}{\Pi(n_p!)}\Pi\{f(x_p, a, b, \dots)\,dx_p\}^{n_p}$$

the products being taken for all values of p.

The 'method of maximum likelihood' consists in choosing values of a, b, ... which make this chance a maximum. As a, b, ... are only involved in f, we have to make

$$\Pi\{f(x_p, a, b, \dots)\,dx_p\}^{n_p}$$

or

$$S\{n_p \log f(x_p, a, b, \dots)\}$$

a maximum.

The difficulty of using the method in general curve fitting is the same as that mentioned in connection with least squares; the equations reached cannot be solved directly and the constants have to be found by approximation.

* E.g. D. Brunt's *Combination of Observations*. Cambridge, 1917.

† Thiele's method of half invariants (*Theory of Observations*) is allied to the method of moments rather than to the method of least squares.

‡ See *Philos. Trans.* A, CCXXII, 309 et seq.; *Proc. Camb. Phil. Soc.* XXII, 700 et seq.

Fisher's papers deal largely with other uses of 'likelihood'. For instance, he uses it to test whether the method of moments will give 'efficient' * values for the constants and throws doubt on the method when the curve is far removed from the 'Normal curve of error' type. The examples given in Chapters 5 and 7 above include curves that may be considered open to this technical criticism. It would not appear, however, that the graduations are bad. Moreover, many other similar curves have been fitted by moments and good practical results have been obtained. As tentative suggestions, perhaps the explanation may be that larger samples than those ordinarily obtained for curve fitting are needed before the tendency to normality is reached, or that the constants are not independent but correlated.† At any rate, until some other better curve fitting process is available, there can be no harm in using a method that, in practice, fits curves satisfactorily to statistical facts. The 'large sample' nature of the 'efficiency' of maximum likelihood estimators is a reason for caution in their use. This point has been emphasized in a letter to *The American Statistician* (XXI (1967), 52) by L. R. Shenton, who also has some kind words for the method of moments.

Minimum χ^2

This method was suggested by Dr Kirstine Smith ‡ and fits curves by making χ^2 a minimum. This comes to the same thing as making $S(n^s/m_s)$ a minimum, where n is the statistical frequency and m the graduated.

Difficulty arises as in the methods described above because the resulting equations cannot be solved directly, and K. Smith suggested that the constants should first be calculated by moments and then adjusted by minimum χ^2. The additional work involved in any such double system of curve fitting would be justified if the improvement in fitting is considerable or if the problem demands refined treatment. The paper quoted gave some examples.

There is a modification of the minimum χ^2 method that is eminently

* The criterion of 'efficiency' is 'that in large samples, when the distributions of the statistics tend to normality, that statistic is to be chosen that has the least probable error'. Fisher uses 'statistic' for a measure obtained from statistical samples: its plural is intended in the tenth word of the quotation.

† Anyone who does a large amount of curve fitting will have noticed that a mistake in working out one constant is not always wholly reflected in the curve. A simple example of what is meant can be inferred from the adjustment for J-shaped curves given in Appendix 1.

‡ *Biometrika*, XI, 262 et seq.

practical. In a study of *Mortality Variations in Sweden*, H. Cramèr and H. Wold * assumed Makeham's law of mortality, † $\mu_x = \alpha + \beta c^x$, and used

$$\chi_1^2 = \Sigma \frac{(\theta_x - \alpha E_x - \beta c^x E_x)^2}{\theta_x}$$

instead of

$$\chi^2 = \Sigma \frac{(\theta_x - E_x \mu_x)^2}{E_x \mu_x}$$

where E_x stands for the 'Exposed to risk' of death at age x and θ_x for the corresponding deaths. Even with the modification, the equations are awkward. Minimization was effected in two stages. The first step was to assume definite value for c, and it was noticed that the methods of least squares and moments gave values for α and β which were equally good when tested by χ^2. Trials were then made with a few values of c and the value of c finally chosen was such as to make χ^2 a minimum on the basis of passing a parabola through the values of χ^2. The following table gives an example of this part of the work:

1911–15. *Males. Values of χ^2, by method of moments. Graduation by parabolas of second and third degree*

$10^3 \log c$	χ^2	Second degree parabola	Third degree parabola
43·25	40·02	39·968	40·020
43·50	32·09	32·143	32·091
43·75	27·41	27·461	27·409
44·00	25·92	25·921	25·921
44·25	27·58	27·525	27·577
44·50	32·32	32·271	32·323
44·75	40·11	40·111	40·109

The value of 44·00 was used for $10^3 \log c$ in this case. Incidentally it will be noticed from the table how great a change in χ^2 is caused by a comparatively small change in the value of log c.

Having then decided on c, new values of β and α were calculated in that order and the α values by moments and by minimum χ^2 are shown.

* *Skandinavisk Aktuarietidskrift*, 1935, pp. 161 et seq.

† Non-actuarial readers must not confuse the force of mortality, represented by μ_x in the international actuarial notation, with a moment.

As described, the method seems lengthy but it probably lent itself to systemization because graduations were in fact made from census particulars for quinquennia from 1801–5 to 1926–30 and for quinquennial ages from 32·5 to 87·5 for both sexes. The results were cross graduated so as to produce a smooth mortality surface for forecasting mortality. Similar tables were also produced for generations.

As an example of graduation on a large scale the paper should be studied both for its practical importance and interesting methods.

METHOD OF PERCENTILE POINTS

This method consists of equating theoretical and observed values for a number of percentile points. It has been used in Chapter 7 and is convenient for the curves described in that chapter and in some other cases. It is not, in general, suitable for fitting Pearson curves of the main Types.

APPENDIX 6

KEY TO THE ACTUARIAL TERMS AND SYMBOLS USED

The following explanation of certain technical terms and symbols that are used in this book is given as an assistance to non-actuarial readers. For a fuller account of the functions and notation, reference can be made to the text-books for actuaries, *Compound Interest and Annuities-Certain* by D. W. A. Donald and *Life and Other Contingencies* by P. F. Hooker and L. H. Longley-Cook. They are published by the Cambridge University Press.

When an investigation is made into the mortality experienced among lives assured, the number of persons entering at each age, and the numbers passing out of observation at each age owing to (1) *death*, (2) *withdrawal*, by the policies lapsing, being surrendered or terminating from some other cause, are recorded. The *exposed to risk* of death at age 25 (E_{25}) means the number who had the chance of dying between ages 25 and 26, and were on the average at risk for the whole year.

The number who die between ages 25 and 26, divided by the exposed to risk at age 25, gives *the probability of dying in a year* at age 25 (q_{25}). Consequently, deaths are tabulated according to age last birthday when the exposed to risk are tabulated according to the exact age or an approximation thereto; if, as is often necessary, the investigation is made according to the number of years an insurance has been in force, the exposed to risk are tabulated according to an exact duration and deaths according to the integral number of years in force (*curtate duration*).

When an experience ends in any year, say 1930, there will be a large number of persons who have been at risk, but whose policies are still in force; these are called *existing at the close of the observation.* When a graduation has been made the *expected deaths* are found by multiplying the exposed to risk by the graduated values of q_x. The result is then compared with the actual deaths.

$O^{NM(5)}$ is the name given to the table of mortality obtained from the male lives assured by ordinary whole-life without profit policies between 1863 and 1893, excluding the first five years of assurance.

O^{M} is a table constructed from the similar with-profit assurances for all durations, and H^M (healthy males) is the name given to the older experience which ended in 1863.

q_x is (see above) the probability of dying in a year.

p_x is the probability of a person aged x living one year. So if we imagine a stationary community, which a person can only leave by death, and consider l_x to be the number living at exact age x, then $l_{x+1} = p_x \times l_x$ and $l_{x+2} = p_{x+1} \times l_{x+1}$, and so on.

The value, at a rate of interest i per unit, of a sum of 1 payable if a person aged x be alive at the end of n years, is therefore $v^n l_{x+n}/l_x$ where $v = (1+i)^{-1}$, and the value of an annuity of 1 would be

$$\frac{vl_{x+1}+v^2l_{x+2}+\dots}{l_x}$$

Now for convenience in making tables it is well to multiply numerator and denominator of this expression by v^x, and we have as the value of the annuity

$$\frac{v^{x+1}l_{x+1}+v^{x+2}l_{x+2}+\dots}{v^x l_x} = \frac{D_{x+1}+D_{x+2}+\dots}{D_x} = \frac{N_{x+1}}{D_x}$$

where $D_x = v^x l_x$ and $N_x = D_x + D_{x+1} + \dots$

Similarly $S_x = N_x + N_{x+1} + \dots$

Tables of D, N, and S are called *commutation* columns.

$a_{\overline{n}|}$ is written for the value of an annuity of 1 payable for n years certain, independent of any life, so its value is

$$v+v^2+\dots+v^n$$

$\bar{a}_{\overline{n}|}$ is the value of a similar annuity of 1 per annum payable j times a year when j takes the limit of ∞, so that its value is

$$\int_0^n v^m \, dm$$

Colog p_x is the logarithm of the reciprocal of p_x, and Makeham's hypothesis assumes that its value is $A+Bc^x$; but

$$\text{colog}\, p_x = \log l_x - \log l_{x+1};$$

therefore an alternative way of stating the hypothesis is $l_x = ks^x g^{c^x}$.

The *force of mortality* (μ_x) is $-\frac{1}{l_x}\frac{dl_x}{dx}$. On Makeham's hypothesis it takes the form $\alpha+\beta c^x$.

When valuing the policies in an assurance office the actuary groups cases together to save labour. When assurances are payable at death they are grouped according to the year of birth, but when they are payable at a certain *maturity age* or previous death (*Endowment Assurances*) they can be grouped either according to year of birth or according to the number of years to run (*unexpired term*). In the latter case they are usually valued by finding an average age at maturity; formerly this was done by taking the mean of the ages, but G. J. Lidstone has shown that a more accurate result is reached by weighting the ages in Geometrical Progression. The constants used for this purpose are called *Z*.

A *model office* is an imaginary specimen office which is used for making approximate valuations.

APPENDIX 7

ABRIDGED READING

A student who needs to know little of the details of curve fitting and graduation but wants to have a general idea of what moments and frequency curves mean, should read Chapters 1 and 2 and the first eight articles of Chapter 3, introducing the subject of fitting curves by moments (§§ 9–13 give an alternative arithmetical method and may be omitted). The student should understand that the arithmetical calculations of moments give a result that approximates to the corresponding moments found from the curve adopted for graduation (see §§ 15 and 16), but §§ 17–22 may be omitted. §§ 23–5 may be read, but little time need be spent on the algebra, and the adjustments of § 26 may be accepted as sufficient. § 27 is a general summary and should be read carefully.

Chapter 4 may be abridged by reading § 1 and examining the table on page 45 (especially the 'Remarks' column) and the diagram at the end of the book. These curves form Pearson's complete system: they express the frequency (y) and are evolved from the differential equation

$$\frac{d \log y}{dx} = \frac{x+a}{b_0+b_1x+b_2x^2}$$

Chapter 5 shows how each type can be fitted by the method of moments. All the mathematical part may be omitted, but the numerical examples distributed through the chapter should be examined in order to form a proper idea of the range of applicability of the system of curves. Pages 70–74 should be read; they deal with the 'normal curve of error' and supply a simple example of curve fitting.

Chapters 6, 7 and 8 can be studied in a similar way to Chapter 5, without detailed examination of the mathematics. The remaining chapters (and appendices) are intended for quick reference as need arises. It is not essential (though it is helpful) to read them to comprehend the curve-fitting methods described earlier in this book. However, it is desirable to study Appendix 5.

14

APPENDIX 8

REFERENCES AND BIBLIOGRAPHY

For working out examples similar to those in this book, it is desirable to use a 7-figure table of logarithms, trigonometric functions, etc., such as *Chambers'* and Barlow's *Tables of Squares*, etc. A multiplying machine will save much heavy arithmetic.

The following will be found of help in various ways:

Biometrika Tables for Statisticians. Edited by E. S. Pearson and H. O. Hartley, Camb. Univ. Press.

Tables for Statisticians and Biometricians, Parts I and II. Edited by K. Pearson. Camb. Univ. Press.

Tracts for Computers. Camb. Univ. Press, especially IX, giving $\log \Gamma(x)$ from $x = 1$ to 50·9 by intervals of 0·01; XV and XXIV, *Random Sampling Numbers*, and XXIII, *Tables of* $\tan^{-1} x$ *and* $\log(1+x^2)$.

The following lists of books and papers deal with the subject generally or with parts of it. The lists are confined to work written in English and do not aim at a complete bibliography. The books in the first list deal with parts of the subject from various points of view and will be found useful for further reading. The lists of papers include those mentioned in the text but do not give all the authorities consulted. Papers are not always included when their contents are more conveniently read in one of the books (e.g. in Fisher's *Statistical Methods* or in the Introductions to various statistical tables). The division between subjects is necessarily somewhat arbitrary.

Most of the original work in English on statistical mathematics has appeared in the *Philosophical Transactions* or the *Proceedings of the Royal Society*, the *Transactions of the Royal Society of Edinburgh*, or in such journals as *Biometrika, Metron, Journal of Royal Statistical Society, Journal of the American Statistical Association, Technometrics, Biometrics, Sankhyā. Annals of Eugenics, Annals of Mathematical Statistics* and on actuarial subjects in the *Journals* of the Institute of Actuaries and the Faculty of Actuaries or in *Skandinavisk Aktuarietidskrift.* A student who wishes to trace the past development of the subject or keep in touch with future advances may be referred to them.

LIST OF BOOKS

A. L. Bowley. *F. Y. Edgeworth's Contributions to Mathematical Statistics.* Royal Statistical Society.

D. Brunt. *The Combination of Observations.* Camb. Univ. Press.

R. A. Fisher. *Statistical Methods for Research Workers.* Oliver & Boyd. *The Design of Experiments.* Oliver & Boyd.

D. C. Jones. *First Course in Statistics.* G. Bell & Sons.

T. L. Kelley. *Statistical Method.* The Macmillan Co.

E. S. Pearson. *Application of Statistical Methods to Industrial Standardisation and Quality Control.* British Standards Institution.

H. E. Soper. *Frequency Arrays.* L. Reeve & Co., Ltd.

J. F. Steffensen. *Some Recent Researches in the Theory of Statistics and Actuarial Science.* Camb. Univ. Press.

L. H. C. Tippett. *Methods of Statistics.* Williams & Norgate.

E. T. Whittaker and G. Robinson. *Calculus of Observations.* Blackie & Co.

G. U. Yule and M. G. Kendall. *Introduction to the Theory of Statistics.* Charles Griffin & Co.

LISTS OF PAPERS

Methods of Moments, etc.

H. Cramèr and H. Wold. 'Mortality Variations in Sweden.' *Skand. Aktuarietidskrift* (1935), 161.

R. A. Fisher. 'On the mathematical foundations of theoretical statistics.' *Philos. Trans. A,* CCII, 309. See also *Proc. Camb. Phil. Soc.* XXII, 700.

E. S. Martin. 'On corrections for the moment coefficients of frequency distributions when the start of the frequency is one of the characteristics to be determined.' *Biometrika,* XXVI, 12.

E. Pairman and K. Pearson. 'Corrections for moment coefficients.' *Biometrika,* XII, 231.

E. S. Pearson. 'Some problems arising in approximating to probability distributions, using moments.' *Biometrika* L (1963), 95.

K. Pearson. 'Systematic fitting of curves to observations.' *Biometrika,* I, 265; II, 1.

W. F. Sheppard. 'Calculation of the most probable values of frequency constants, etc.' *Proc. Lond. Math. Soc.* XXIX, 353.

Frequency Curves

J. Aitchison and J. A. C. Brown, *The Lognormal Distribution.* Camb. Univ. Press, 1957.

C. V. L. Charlier. 'Researches into the theory of probability.' *Meddelanden Lunds Astronomiska Observatorium,* 1906. 'A new form of the Frequency Function.' Reprinted, 1928.

H. Cramèr. 'On the composition of elementary errors.' *Skand. Aktuarietidskrift* (1928), pp. 13 and 141.

F. Y. Edgeworth. 'Mathematical representation of statistical data.' *J. Roy. Statist. Soc.* LXI, 691; LXIX, 497; LXX, 102; LXXIX, 455; LXXX, 65, 266, 411; LXXXVII, 571. 'Generalised Law of Error.' *Proc. Camb. Phil. Soc.* XX, 36, 113.

W. P. Elderton. 'An approximate law of survivorship and other notes on the use of frequency curves in actuarial statistics.' *J. Inst. Actu.* LXV, 1.

N. L. Johnson, 'Systems of frequency curves generated by methods of translation.' *Biometrika*, XXXVI (1949), 149.

N. L. Johnson. 'Tables to facilitate fitting S_U curves.' *Biometrika*, LII (1965), 547.

N. L. Johnson, E. Nixon, D. E. Amos and E. S. Pearson. 'Table of percentage points of Pearson curves, for given $\sqrt{\beta_1}$ and β_2, expressed in standard measure.' *Biometrika*, L (1963), 459.

J. C. Kapteyn. *Skew Frequency Curves in Biology and Statistics*. Groningen, 1903 and 1916.

E. S. Pearson and J. W. Tukey. 'Approximate means and standard deviations based on distances between percentage points of frequency curves.' *Biometrika*, LII (1965), 533.

K. Pearson. 'Skew variation in homogeneous material.' *Philos. Trans.* A, CLXXXVI, 343; CXCVII, 443; CCXVI, 429.

E. C. Rhodes. 'On the generalised law of error.' *J. Roy. Statist. Soc.* LXXXVIII, 576.

Frequency Surfaces

C. V. L. Charlier. 'Contributions to the mathematical theory statistics.' 6. The correlation function of Type A, *Ark. Mat. Astr. Fys.* IX, no. 26 (1914), 1.

F. Y. Edgeworth. 'On the mathematical representation of statistical data. *J. R. Statist. Soc.* LXXX (1917), 266.

N. L. Johnson. 'Bivariate distributions based on simple translation systems.' *Biometrika*, XXXVI (1949), 297.

K. Pearson. 'Notes on skew frequency surfaces.' *Biometrika*, XV (1923), 222.

K. Pearson. 'The fifteen constant bivariate frequency surface.' *Biometrika*, XVII (1925), 268.

S. J. Pretorius. 'Skew bivariate frequency surfaces, examined in the light of numerical illustrations.' *Biometrika*, XXII (1930), 109.

E. C. Rhodes. 'On a certain skew correlation surface.' *Biometrika*, XIV (1923), 355.

E. C. Rhodes. 'On a skew correlation surface.' *Biometrika*, XVII (1925), 314.

M. J. Van Uven. 'Extension of Pearson's probability distributions to two variables, I–IV.' *Proc. Kon. Akad. Wetens., Amsterdam*, L (1947), 1063, 1252; LI (1948), 41, 191.

S. D. Wicksell. 'The application of solid hypergeometrical series to frequency distributions in space.' *Phil. Mag., 6th Series*, XXXIV (1917), 389.

Standard Errors, Goodness of Fit, etc.

D. E. Barton. 'On Neyman's smooth test of goodness of fit and its power with respect to a particular system of alternatives.' *Skand. Aktuar.* XXXVI (1953), 24.

D. E. Barton. 'A form of Neyman's Ψ^2 test of goodness of fit applicable to grouped and discrete data.' *Skand. Aktuar.* XXXVIII (1955), 1.

Biometrika, editorial. 'Probable errors of frequency constants.' *Biometrika*, II, 273.

F. N. David. 'A χ^2-smooth test for goodness of fit.' *Biometrika*, XXXIV (1947), 299.

R. A. Fisher. Various papers dealing with the subject and especially with χ^2 and small samples. *J. Roy. Statist. Soc.* LXXXV, 597; LXXXVII, 442; *Economica* (1923), p. 139; *Biometrika*, X, 507; *Metron*, I, Pt IV, 1; V, Pt III, 92.

R. Henderson. 'Frequency curves and moments.' *Trans. Actu. Soc. Amer.*, VIII, 30 or *J. Inst. Actu.* XLI, 429.

M. Greenwood. 'Errors in random sampling.' *Biometrika*, IX, 69.

J. Neyman. ' "Smooth" test for goodness of fit.' *Skand. Aktuar.* XX (1937), 150.

E. S. Pearson. 'Some notes on sampling tests with two variables.' *Biometrika*, XXI, 337.

K. Pearson. 'Criterion that a given system of deviations can have arisen from Random Sampling.' *Phil. Mag.* July 1900. See also various other papers. *Biometrika*, VIII, 250; XI, 292; XIV, 186, 418. (With L. N. G. Filon.) *Philos. Trans.* A, CXCI, 229.

'Student.' 'Small samples.' *Biometrika*, VI, 302; X, 384; XV, 271. 'Ranks and grades correlation method.' *Biometrika*, XIII, 264.

APPENDIX 9

TABLES

Useful Constants

$e =$	$2{\cdot}71828\ 18285$	$\log(\log_{10} e) =$	$\bar{1}{\cdot}63778\ 43114$
$e^{-1} =$	$0{\cdot}36787\ 94417$	$\log_{10}\pi =$	$0{\cdot}49714\ 98728$
$\pi =$	$3{\cdot}14159\ 26536$	$\log_{10}\sqrt{\pi} =$	$0{\cdot}24857\ 49364$
$\log_{10} e =$	$0{\cdot}43429\ 44820$	$\log_{10}\dfrac{1}{\sqrt{(2\pi)}} =$	$\bar{1}{\cdot}60091\ 00657$
$\log_e 10 =$	$2{\cdot}30258\ 50930$	$\log_{10} e^{-\frac{1}{12}} =$	$\bar{1}{\cdot}96380\ 87932$

Areas and ordinates of Normal Curve of Error in terms of abscissa

x	$P(x)$	$Z(x)$
0·0	0·5000	0·3989
0·1	0·5398	0·3970
0·2	0·5793	0·3910
0·3	0·6179	0·3814
0·4	0·6554	0·3683
0·5	0·6915	0·3521
0·6	0·7257	0·3332
0·7	0·7580	0·3123
0·8	0·7881	0·2897
0·9	0·8159	0·2661
1·0	0·8413	0·2420
1·1	0·8643	0·2179
1·2	0·8849	0·1942
1·3	0·9032	0·1714
1·4	0·9192	0·1497
1·5	0·9332	0·1295
1·6	0·9452	0·1109
1·7	0·9554	0·0940
1·8	0·9641	0·0790
1·9	0·9713	0·0656
2·0	0·9772	0·0540
2·5	0·9938	0·0175
3·0	0·9987	0·0044
3·5	0·9998	0·0009

$Z(x) = \frac{1}{\sqrt{(2\pi)}\sigma} e^{-\frac{1}{2}(x/\sigma)^2}$. The tabulation assumes $\sigma = 1$.

$$P(x) = \int_{-\infty}^{x} Z(x)\, dx$$

χ^2 *Test for Goodness of Fit. Values of P*

χ^2	$\nu = 2$	$\nu = 5$	$\nu = 8$	$\nu = 11$	$\nu = 14$	$\nu = 17$	$\nu = 20$	$\nu = 23$	$\nu = 26$	$\nu = 29$	χ^2
3	0·223	0·700	0·934	0·991	0·999	1·000	1·000	1·000	1·000	1·000	3
6	0·050	0·306	0·647	0·873	0·966	0·993	0·999	1·000	1·000	1·000	6
9	0·011	0·109	0·342	0·622	0·831	0·940	0·983	0·996	0·999	1·000	9
12	0·002	0·035	0·151	0·363	0·606	0·800	0·916	0·970	0·991	0·998	12
15	0·001	0·010	0·059	0·182	0·378	0·595	0·776	0·895	0·957	0·985	15
18	0·000	0·003	0·021	0·082	0·207	0·389	0·589	0·757	0·876	0·944	18
21	0·000	0·001	0·007	0·033	0·102	0·226	0·397	0·581	0·742	0·859	21
24	0·000	0·000	0·002	0·013	0·046	0·119	0·243	0·404	0·576	0·729	24
27	0·000	0·000	0·001	0·005	0·019	0·058	0·135	0·256	0·409	0·572	27
30	0·000	0·000	0·000	0·002	0·008	0·026	0·070	0·149	0·268	0·414	30

If $\chi^2 = 0$, $P = 1$. Rough interpolations can be made either horizontally or vertically, but not diagonally, by using log P instead of P. ν is the number of 'degrees of freedom'.

Table of log $\Gamma(p)$

p	0		1	2	3	4	5	6	7	8	9
1·00	...	...	9750	9500	9251	9003	8755	8509	8263	8017	7773
1·01	$\bar{1}$·99	7529	7285	7043	6801	6560	6320	6080	5841	5602	5365
1·02	$\bar{1}$·99	5128	4892	4656	4421	4187	3953	3721	3489	3257	3026
1·03	$\bar{1}$·99	2796	2567	2338	2110	1883	1656	1430	1205	0981	0757
1·04	$\bar{1}$·99	0533	0311	0089	$\bar{9}$868	$\bar{9}$647	$\bar{9}$427	$\bar{9}$208	$\bar{8}$989	$\bar{8}$772	$\bar{8}$554
1·05	$\bar{1}$·98	8338	8122	7907	7692	7478	7265	7053	6841	6629	6419
1·06	$\bar{1}$·98	6209	6000	5791	5583	5376	5169	4963	4758	4553	4349
1·07	$\bar{1}$·98	4145	3943	3741	3539	3338	3138	2939	2740	2541	2344
1·08	$\bar{1}$·98	2147	1951	1755	1560	1365	1172	0978	0786	0594	0403
1·09	$\bar{1}$·98	0212	0022	$\bar{9}$833	$\bar{9}$644	$\bar{9}$456	$\bar{9}$269	$\bar{9}$082	$\bar{8}$896	$\bar{8}$710	$\bar{8}$525
1·10	$\bar{1}$·97	8341	8157	7974	7791	7610	7428	7248	7068	6888	6709
1·11	$\bar{1}$·97	6531	6354	6177	6000	5825	5650	5475	5301	5128	4955
1·12	$\bar{1}$·97	4783	4612	4441	4271	4101	3932	3764	3596	3429	3262
1·13	$\bar{1}$·97	3096	2931	2766	2602	2438	2275	2113	1951	1790	1629
1·14	$\bar{1}$·97	1469	1309	1150	0992	0835	0677	0521	0365	0210	0055
1·15	$\bar{1}$·96	9901	9747	9594	9442	9290	9139	8988	8838	8688	8539
1·16	$\bar{1}$·96	8390	8243	8096	7949	7803	7658	7513	7369	7225	7082
1·17	$\bar{1}$·96	6939	6797	6655	6514	6374	6234	6095	5957	5818	5681
1·18	$\bar{1}$·96	5544	5408	5272	5137	5002	4868	4734	4601	4469	4337
1·19	$\bar{1}$·96	4205	4075	3944	3815	3686	3557	3429	3302	3175	3048
1·20	$\bar{1}$·96	2922	2797	2672	2548	2425	2302	2179	2057	1936	1815
1·21	$\bar{1}$·96	1695	1575	1456	1337	1219	1101	0984	0867	0751	0636
1·22	$\bar{1}$·96	0521	0407	0293	0180	0067	$\bar{9}$955	$\bar{9}$843	$\bar{9}$732	$\bar{9}$621	$\bar{9}$511
1·23	$\bar{1}$·95	9401	9292	9184	9076	8968	8861	8755	8649	8544	8439
1·24	$\bar{1}$·95	8335	8231	8128	8025	7923	7821	7720	7620	7520	7420
1·25	$\bar{1}$·95	7321	7223	7125	7027	6930	6834	6738	6642	6547	6453
1·26	$\bar{1}$·95	6359	6267	6173	6081	5989	5898	5807	5716	5627	5537
1·27	$\bar{1}$·95	5449	5360	5273	5185	5099	5013	4927	4842	4757	4673
1·28	$\bar{1}$·95	4589	4506	4423	4341	4259	4178	4097	4017	3938	3858
1·29	$\bar{1}$·95	3780	3702	3624	3547	3470	3394	3318	3243	3168	3094
1·30	$\bar{1}$·95	3020	2947	2874	2802	2730	2659	2588	2518	2448	2379
1·31	$\bar{1}$·95	2310	2242	2174	2106	2040	1973	1907	1842	1777	1712
1·32	$\bar{1}$·95	1648	1585	1522	1459	1397	1336	1275	1214	1154	1094
1·33	$\bar{1}$·95	1035	0977	0918	0861	0803	0747	0690	0634	0579	0524
1·34	$\bar{1}$·95	0470	0416	0362	0309	0257	0205	0153	0102	0051	0001
1·35	$\bar{1}$·94	9951	9902	9853	9805	9757	9710	9663	9617	9571	9525
1·36	$\bar{1}$·94	9480	9435	9391	9348	9304	9262	9219	9178	9136	9095
1·37	$\bar{1}$·94	9054	9015	8975	8936	8898	8859	8822	8785	8748	8711
1·38	$\bar{1}$·94	8676	8640	8605	8571	8537	8503	8470	8437	8405	8373
1·39	$\bar{1}$·94	8342	8311	8280	8250	8221	8192	8163	8135	8107	8080
1·40	$\bar{1}$·94	8053	8026	8000	7975	7950	7925	7901	7877	7854	7831
1·41	$\bar{1}$·94	7808	7786	7765	7744	7723	7703	7683	7664	7645	7626
1·42	$\bar{1}$·94	7608	7590	7573	7556	7540	7524	7509	7494	7479	7465
1·43	$\bar{1}$·94	7451	7438	7425	7413	7401	7389	7378	7368	7358	7348
1·44	$\bar{1}$·94	7338	7329	7321	7312	7305	7298	7291	7284	7278	7273
1·45	$\bar{1}$·94	7268	7263	7259	7255	7251	7248	7246	7244	7242	7241
1·46	$\bar{1}$·94	7240	7239	7239	7240	7241	7242	7243	7245	7248	7251
1·47	$\bar{1}$·94	7254	7258	7262	7266	7271	7277	7282	7289	7295	7302
1·48	$\bar{1}$·94	7310	7317	7326	7334	7343	7353	7363	7373	7384	7395
1·49	$\bar{1}$·94	7407	7419	7431	7444	7457	7471	7485	7499	7514	7529
p	0		1	2	3	4	5	6	7	8	9

Table of log *Γ(p)—continued*

p	0		1	2	3	4	5	6	7	8	9
1·50	$\bar{1}$·94	7545	7561	7577	7594	7612	7629	7647	7666	7685	7704
1·51	$\bar{1}$·94	7724	7744	7764	7785	7806	7828	7850	7873	7896	7919
1·52	$\bar{1}$·94	7943	7967	7991	8016	8041	8067	8093	8120	8146	8174
1·53	$\bar{1}$·94	8201	8229	8258	8287	8316	8346	8376	8406	8437	8468
1·54	$\bar{1}$·94	8500	8532	8564	8597	8630	8664	8698	8732	8767	8802
1·55	$\bar{1}$·94	8837	8873	8910	8946	8983	9021	9059	9097	9135	9174
1·56	$\bar{1}$·94	9214	9254	9294	9334	9375	9417	9458	9500	9543	9586
1·57	$\bar{1}$·94	9629	9672	9716	9761	9806	9851	9896	9942	9989	$\bar{0}$035
1·58	$\bar{1}$·95	0082	0130	0177	0225	0274	0323	0372	0422	0472	0522
1·59	$\bar{1}$·95	0573	0624	0676	0728	0780	0833	0886	0939	0993	1047
1·60	$\bar{1}$·95	1102	1157	1212	1268	1324	1380	1437	1494	1552	1610
1·61	$\bar{1}$·95	1668	1727	1786	1845	1905	1965	2025	2086	2147	2209
1·62	$\bar{1}$·95	2271	2333	2396	2459	2522	2586	2650	2715	2780	2845
1·63	$\bar{1}$·95	2911	2977	3043	3110	3177	3244	3312	3380	3449	3517
1·64	$\bar{1}$·95	3587	3656	3726	3797	3867	3938	4010	4081	4154	4226
1·65	$\bar{1}$·95	4299	4372	4446	4519	4594	4668	4743	4819	4894	4970
1·66	$\bar{1}$·95	5047	5124	5201	5278	5356	5434	5513	5592	5671	5740
1·67	$\bar{1}$·95	5830	5911	5991	6072	6154	6235	6317	6400	6482	6566
1·68	$\bar{1}$·95	6649	6733	6817	6901	6986	7072	7157	7243	7322	7416
1·69	$\bar{1}$·95	7503	7590	7678	7766	7854	7943	8032	8122	8211	8301
1·70	$\bar{1}$·95	8391	8482	8573	8664	8756	8848	8941	9034	9127	9220
1·71	$\bar{1}$·95	9314	9409	9502	9598	9693	9788	9884	9980	$\bar{0}$077	$\bar{0}$174
1·72	$\bar{1}$·96	0271	0369	0467	0565	0664	0763	0862	0961	1061	1162
1·73	$\bar{1}$·96	1262	1363	1464	1566	1668	1770	1873	1976	2079	2183
1·74	$\bar{1}$·96	2287	2391	2496	2601	2706	2812	2918	3024	3131	3238
1·75	$\bar{1}$·96	3345	3453	3561	3669	3778	3887	3996	4105	4215	4326
1·76	$\bar{1}$·96	4436	4547	4659	4770	4882	4994	5107	5220	5333	5447
1·77	$\bar{1}$·96	5561	5675	5789	5904	6019	6135	6251	6367	6484	6600
1·78	$\bar{1}$·96	6718	6835	6953	7071	7189	7308	7427	7547	7666	7787
1·79	$\bar{1}$·96	7907	8028	8149	8270	8392	8514	8636	8759	8882	9005
1·80	$\bar{1}$·96	9129	9253	9377	9501	9626	9751	9877	$\bar{0}$003	$\bar{0}$129	$\bar{0}$255
1·81	$\bar{1}$·97	0383	0509	0637	0765	0893	1021	1150	1279	1408	1538
1·82	$\bar{1}$·97	1668	1798	1929	2060	2191	2322	2454	2586	2719	2852
1·83	$\bar{1}$·97	2985	3118	3252	3386	3520	3655	3790	3925	4061	4197
1·84	$\bar{1}$·97	4333	4470	4606	4744	4881	5019	5157	5295	5434	5573
1·85	$\bar{1}$·97	5712	5852	5992	6132	6273	6414	6555	6697	6838	6980
1·86	$\bar{1}$·97	7123	7266	7408	7552	7696	7840	7984	8128	8273	8419
1·87	$\bar{1}$·97	8564	8710	8856	9002	9149	9296	9443	9591	9739	9887
1·88	$\bar{1}$·98	0036	0184	0333	0483	0633	0783	0933	1084	1234	1386
1·89	$\bar{1}$·98	1537	1689	1841	1994	2147	2299	2453	2607	2761	2915
1·90	$\bar{1}$·98	3069	3224	3379	3535	3690	3846	4003	4159	4316	4474
1·91	$\bar{1}$·98	4631	4789	4947	5105	5264	5423	5582	5742	5902	6062
1·92	$\bar{1}$·98	6223	6383	6544	6706	6867	7029	7192	7354	7517	7680
1·93	$\bar{1}$·98	7844	8007	8171	8336	8500	8665	8830	8996	9161	9327
1·94	$\bar{1}$·98	9494	9660	9827	9995	$\bar{0}$162	$\bar{0}$330	$\bar{0}$498	$\bar{0}$666	$\bar{0}$835	$\bar{1}$004
1·95	$\bar{1}$·99	1173	1343	1512	1683	1853	2024	2195	2366	2537	2709
1·96	$\bar{1}$·99	2881	3054	3227	3399	3573	3746	3920	4094	4269	4443
1·97	$\bar{1}$·99	4618	4794	4969	5145	5321	5498	5674	5851	6029	6206
1·98	$\bar{1}$·99	6384	6562	6740	6919	7098	7277	7457	7637	7817	7997
1·99	$\bar{1}$·99	8178	8359	8540	8722	8903	9085	9268	9450	9633	9816
p	0		1	2	3	4	5	6	7	8	9

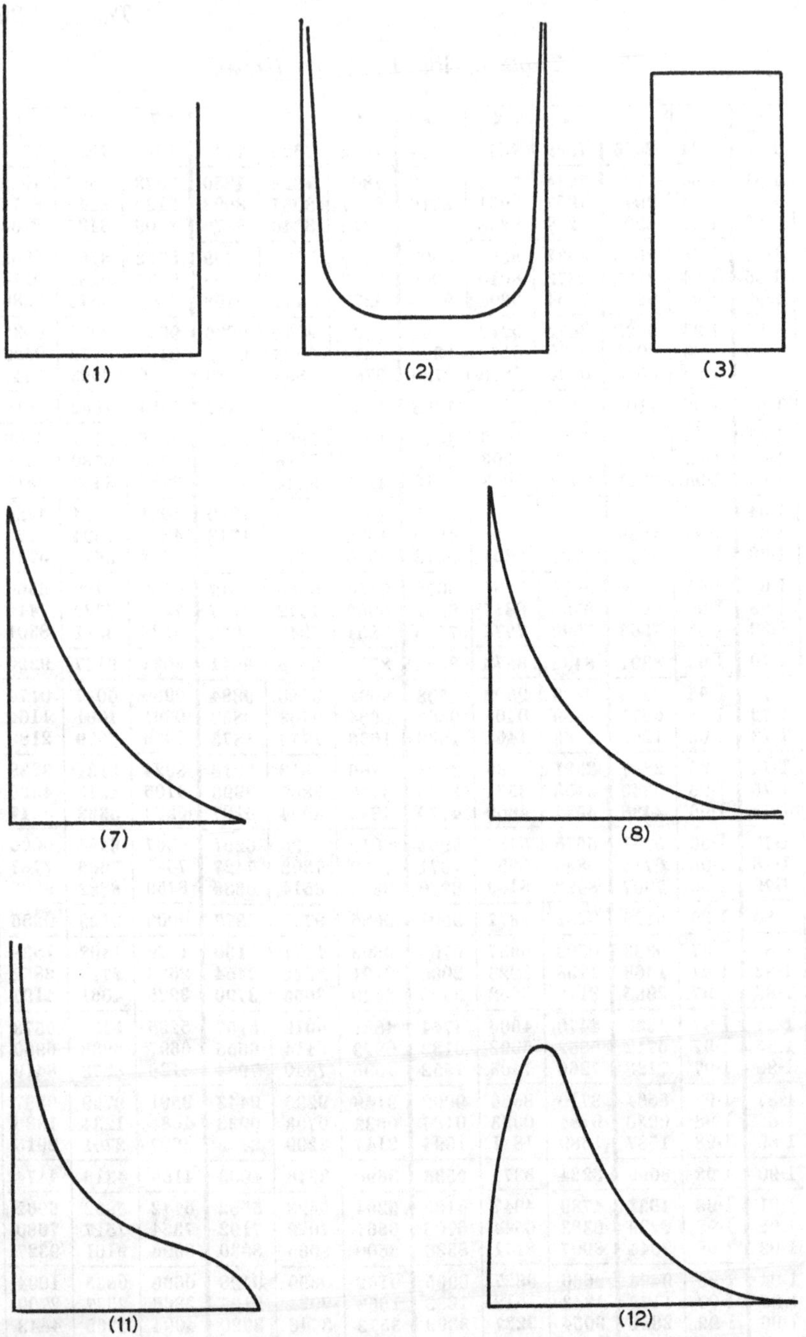

Diagram showing transition from two separated blocks of frequency through U-shaped and J-shaped curves to the more common 'cocked hat' shapes.

(1) represents two separated blocks of frequency developing into U-shaped curve in (2). The horizontal line of (3) is, as it were, the bottom piece of the U-curve and the Type VIII curve of (4) is like part of the U-curve. (5) and (6)

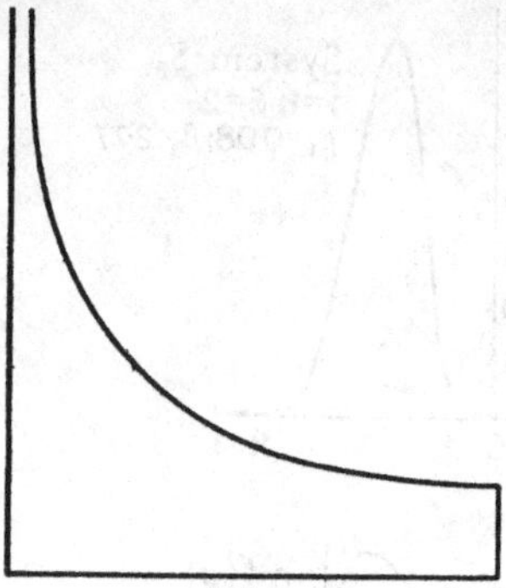

(4)

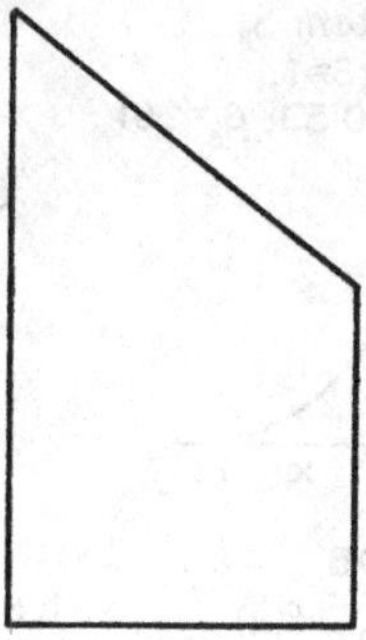

(5)

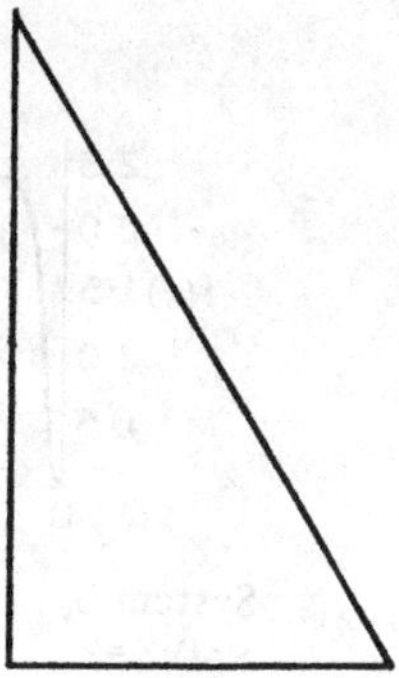

(6)

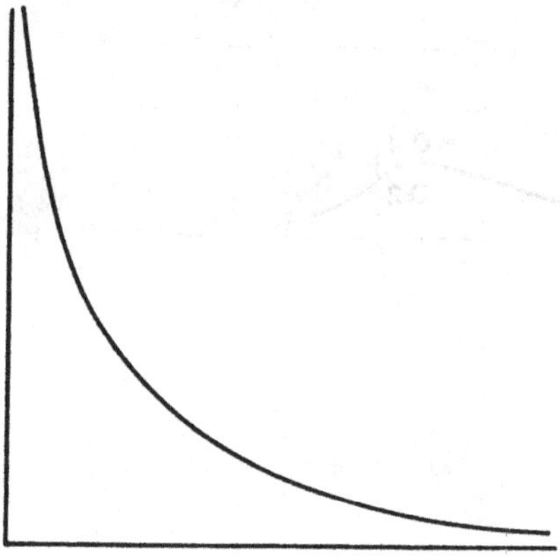

(9)

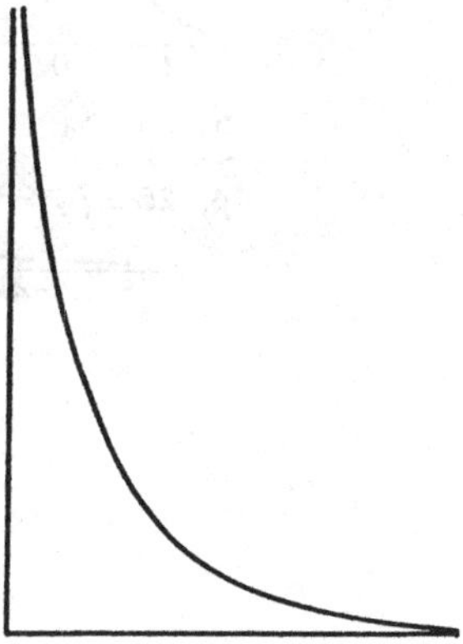

(10)

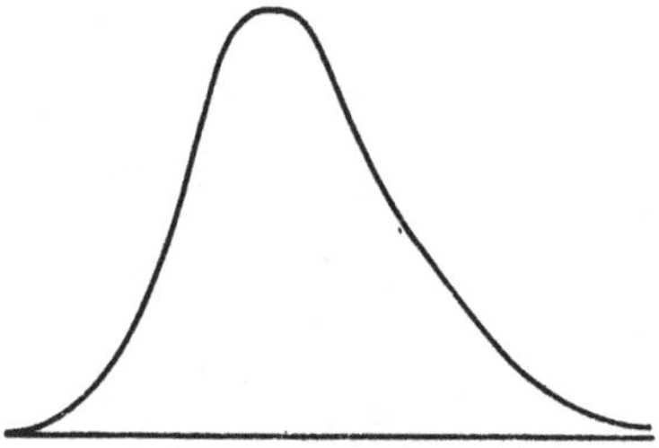

(13)

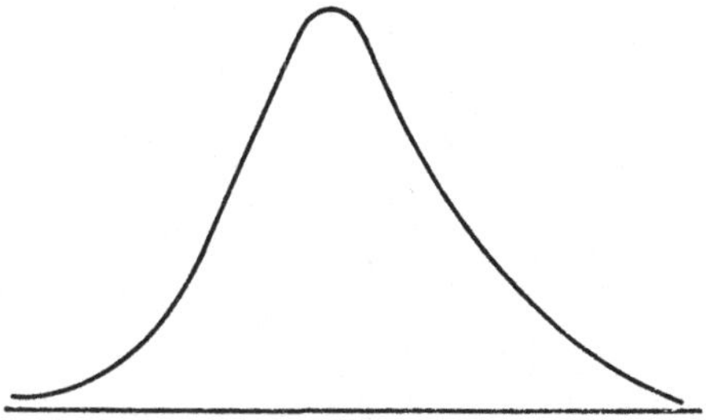

(14)

are limits when straight lines are reached. (7) is Type IX and (8) is the exponential. The next two curves (9) and (10) are **J**-shaped curves of Types III, and I, and (11) is Type (XII). From this we proceed to Types, I, III, IV, V and VI, curves of the 'cocked hat' shape, three examples being given in (12), (13) and (14).

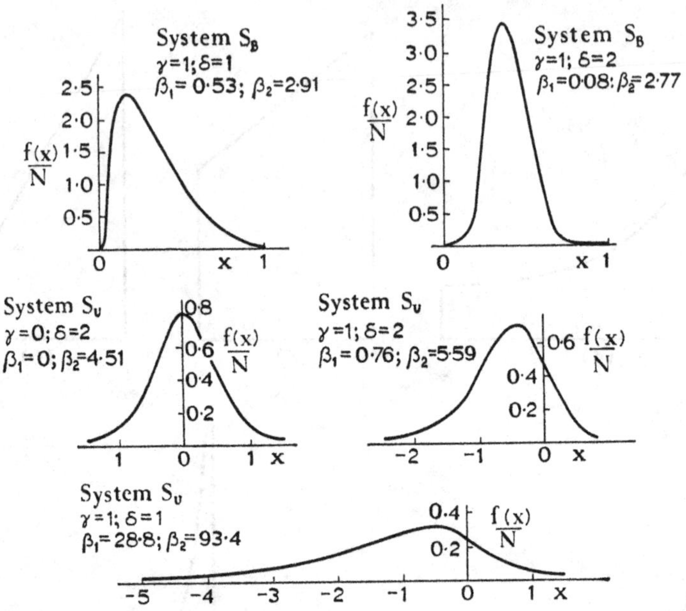
System S_B
$\gamma=1; \delta=1$
$\beta_1=0\cdot53; \beta_2=2\cdot91$
2·5
2·0
$\frac{f(x)}{N}$ 1·5
1·0
0·5
0
x
1
3·5
3·0
2·5
$\frac{f(x)}{N}$ 2·0
1·5
1·0
0·5
0
x 1
System S_B
$\gamma=1; \delta=2$
$\beta_1=0\cdot08; \beta_2=2\cdot77$
System S_U
$\gamma=0; \delta=2$
$\beta_1=0; \beta_2=4\cdot51$
0·8
0·6
0·4
0·2
$\frac{f(x)}{N}$
1
0
1 x
System S_U
$\gamma=1; \delta=2$
$\beta_1=0\cdot76; \beta_2=5\cdot59$
0·6 $\frac{f(x)}{N}$
0·4
0·2
−2
−1
0 x
System S_U
$\gamma=1; \delta=1$
$\beta_1=28\cdot8; \beta_2=93\cdot4$
0·4
0·2
$\frac{f(x)}{N}$
−5
−4
−3
−2
−1
0
1 x

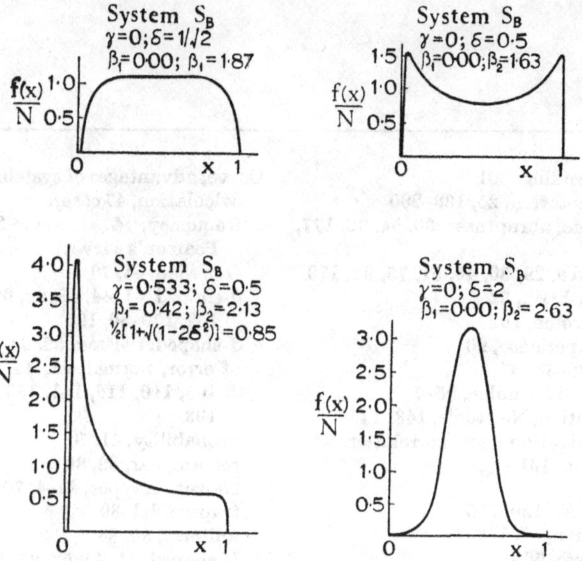

(Note that for S_U, when λ is positive, μ_3 and $\sqrt{\beta_1}$ are negative)

Index

Abridged reading, 201
Actuarial notation, 25, 198–200
Adjustments, abruptness, 30, 54, 92, 177, 180–1
  Sheppard's, 29, 30, 46, 51, 75, 92, 179
Allin, S. H. J. W., 32
Alsace-Lorraine, 162
Annuity experience, 90
  values, 32–3
Approximate formulae, 95–7
Approximation, Newton's, 143, 145
  to distribution of correlation coefficient, 161

Barton, D. E., 120, 176
Beta function, 56, 182–4
  incomplete, 56
Binomial series, 1, 6, 36, 110–12
  standard error in, 156
Bivariate surface, normal, 139–50, 185–7, 188
  Pearson, 137–9
  translation (S_{IJ}), 147–50
  Type AA, 145–7
Block diagram, 5
British Offices Tables, 18
Brownlee, J., 183
Bruns, H., 111
Brunt, D., 194

Carlisle Table, 89
Census, 197
Centroid vertical, 17–18, 54, 65
Charlier, C. V. L., 111, 112, 117
Checking calculations, 15–16
Chi-squared, 167 *et seq.*, 186
  minimum, 195–7
  test, 173–5; table for goodness-of-fit, values of P, 207
  Commutation columns, 199
  Comrie, L. J., 59
  Conditional frequency distribution, 137, 148
  Correlation coefficient, 137, 152–3, 156–7
    estimation of, 141–5
    standard error, 161
    transformation of, 161
  Covariance, 137
  Cramer, H., 115, 196
  Crelle's tables, 100, 102
  Cricket scores, 92
  Criteria for type of Pearson curve, 41
Curve, advantages of systems of, 2
  calculation, 47 *et seq.*
  frequency, 4 *et seq.*, 37 (see also Pearson's curves)
  Gaussian, 41, 70
  J-shaped, 41, 44, 48, 52, 67, 79, 117–19, 162, 179–80, 195
  J-shaped, twisted, 53, 94–5, 180
  of error, normal, 41, 44, 70–4, 76, 78, 103, 110, 115, 141, 153, 166, 188–92, 193
  probability, 41, 70
  rectangular, 86, 88
  transition types, 43–4, 70–95
  trapezoidal, 89
  uniform, 86, 88
  U-shaped, 41, 44, 52, 95, 119, 131

David, F. N., 161, 175
De Moivre, A., 70
Degrees of freedom, 167
Dennis, K. E. R., 120
Deviation from symmetry, see Skewness,
  proportional, 1
  mean, 9–10, 39
  standard, 8–10, 39, 152–4
Differential equation, 35
  of Pearson curves, 37, 201
Dirichlet distributions, 139
Distributions, chance, 110
  exponential, 179–80 (see also Pearson Type X)
  frequency, 4–11
  mean of, 6–8, 10
  mode of, 8, 10
  symmetrical, 60, 75
Distributions of deaths, 4, 82
  entrants, limited payment policies, 67
  existing at close of observation, 5
  exposed to risk, 4, 5, 173
  number of wives, 79
  reserves, 71
  sums assured, 71
  unexpired term, 75
  withdrawals, 5, 108, 134
Donald, D. W. A. 198
Duration, actual, 4
  curate, 4, 5, 198

Edgworth curves (Type Ab), 114, 116, 118, 119–20, 146
Edgeworth, F. Y., 111, 114

Efficiency, 195
Euler-Maclaurin expansion, 177–80
  Type A (Aa or Gram-Charlier), 111–12, 115, 116, 119–21, 146–7, 172
  Type Ab (Edgeworth), 114, 116, 118, 119–20, 146–7
  Type B (Charlier), 112–14, 116, 117, 172
Expected deaths, 198
Exponential distribution, 179–80 (see also Pearson Type X)
Exposed to risk, 196, 198; distributions of, 4, 5, 173

Factorial moments, 21, 22
Fisher, R. A., 161, 194
Force of mortality, 12, 37, 196, 199
Fraser, D. A. S., 175
Freedom, degrees of, 167
Frequency, 10
  conditional, 137, 148
  curves, 4 *et seq.*, 37; further reading, 203 (see also Pearson's curves)
  distribution, 4 *et seq.*
  marginal, 135
  polygons, 4
  surfaces, 135 *et seq.*; further reading, 204; bivariate normal, 139–50, 185–7, 188; Pearson, 137–9; translation (S_{IJ}), 147–50; Type AA, 145–7

Gamma function, 52, 75, 182–4
  incomplete, 56, 183–4
  table of log of, 208–9
Gauss, C. F., 70
Gaussian curve, 41, 70
Geometrical progression, 1, 6, 92, 179, 200
Goodness of fit, 165–76
  further reading, 204
  table for χ^2 test of, 207
Graduation, 2, 12, 24, 48, 55, 71, 76, 89, 92, 94, 173, 196, 198
  formulae for, 13, 35
Gram, J. P., 112
Great numbers, law of, 111
Groups, 2, 14, 102

H^M Table, distribution, 170, 199
Hardy, Sir G. F., 18, 75, 106
Hartley, H. O., 16, 202
Henderson, R., 159
Hermite polynomials, 112
High contact, 27–30, 49, 178
  mathematical definition, 28
High order moments, 39–40

Histogram, 4
Hooker, P. F., 198
Hypergeometric series, 36, 110–11

Incomplete beta function, 56
  gamma function, 56, 183–4

J-shaped curve, 41, 44, 48, 52, 67, 79, 117–19, 162, 179–80, 195
  twisted, 53, 94–5, 180
Johnson, N. L., 123, 173

Kapteyn, J. C., 111
Khamis, S. H., 114
King, G., 82, 170

Laplace, P. S., 70
Law of great numbers, 111
  small numbers, 112
Least squares, method of, 193–4
Lidstone, G. J., 22, 75, 171, 200
Limited payment policies, entrants, 67, 68
Logarithms, use of, 50, 51
Lognormal, 123, 124–6
  line, 124
Longley-Cook, L. H., 198

Macdonnell, W. R., 144
Mackenzie Lees, M., 99
Maclaurin's theorem, 35
Makeham's hypothesis, 12, 103, 196, 199
Marginal curve, 135
Marriage rates, 99
Martin, E. S., 181
Massey, F. J., 176
Maximum likelihood, method of, 194
Mean, 6, 8, 11
  deviation, 9, 39
  standard error of, 157
Median regression, 149
Method of least squares, 193–4
  maximum likelihood, 194–5
  minimum, 2, 197–7
  moments 12 *et seq.*, 34; further reading, 203
  percentile points, 131–3, 197
  summation, 18 *et seq.*, 53, 99
  translation, 115, 122 *et seq.*
Mixture of normal curves, 103–6
Mode, 8
Model office, 200
Moments, 14
  calculation of, 14 *et seq.*, 181
  centroid vertical, about the, 17, 54, 65
  corrections of, 29, 177, 181
  factorial, 21–2

Moments—(contd.)
formulae for, 51
high order, 39–40
inertia, 13
method of, 12 *et seq.*
notation for, 15
*n*th, 14
ordinates, of, 14
summations of, 18 *et seq.*
Mortality, force of, 37, 196, 199
variations in Sweden, 196

Negative frequencies, 120
Newton's approximation method, 143, 145
Neyman, J., 175
Nicholson, C., 141
Normal bivariate surface, 139–50, 185–7, 88
curve of error, 41, 44, 70–4, 76, 78, 103, 110, 115, 141, 153, 166, 188–92, 193; table, 207
mixture, 103–6

OM Table distribution, 170, 198
ONM(5) Table, 100, 105, 198
Owen, D. B. 141

Pairman, E., 177
Pearson curves, differential equation, 37, 201
Type I, 42, 45, 51–8, 95, 102, 106–9, 115, 116, 131, 132, 134, 138, 139, 171–2, 180; II, 43, 45, 74–7, 78, 95, 115, 118, 131, 138, 139; III, 43, 45, 78–81, 98, 99, 100, 114, 116, 120, 138, 139, 171, 178, 183, 184; IV, 42, 45, 58–66, 83, 118, 120, 126, 128, 130, 134, 138; V, 44, 45, 81–5, 171; VI, 43, 45, 67–70, 126, 138, 139; VII, 45, 77–8, 139; VIII, 46, 85–9; IX, 46, 89–91; X, (exponential), 46, 89, 91–2, 179; XI, 46, 92–3; XII, 94
surfaces, 137–9
Pearson, E. S., 16, 95, 134, 202
Pearson, K., 16, 39, 41, 111, 117, 167, 177, 202
Percentile points, method of, 131–3, 197
Poisson series, 112
Polygons, frequency, 4
Probable error, 154
Probability, 1
curve, 41, 70
elementary, 2
Product moment, 136
reduced, 136

Quadrature formulae, 25–6, 28, 47, 76, 102, 108–9

Rates, 98
Rectangular curve, 86, 88
Regression, 137, 138
linear, 138
median, 149
Rhodes, E. C., 120
Runs of signs, 174

Sampling, 162–3
S_B system, 123, 125, 130–34
S_U system, 123, 125, 126–30, 211
compared with Type IV, 130, 134
Seal, H. L., 175
Semi-invariants, 21, 112, 114
Shenton, L. R., 195
Sheppard, W. F., 72
Sheppard's adjustments, 29, 30, 46, 51, 75, 92, 179
formula, 141
Simpson's rule, 47, 55, 76
Skewness, 9–10, 48
of Pearson curves, 39
Small numbers, law of, 112
Smirnov, N., 176
'Smooth' test, 175
Soper, H. E., 167
Statistical tables, 2
Standard deviation, 8–10, 39, 152–4
Standard errors, 151 *et seq.*; further reading, 204
of estimator of $\sqrt{\beta_1}$, 161
of estimator of β_2, 161
mean, 157
Steffensen, J. F., 114–15
Summation of moments, 19–21
Summation method, 18 *et seq.*, 53, 99
Sutton's sickness tables, 60

Tetley, H., 174
Tiku, M. L., 114
Transformation of correlation coefficient, 161
Transition types, 43–4, 70–95
Transition, method of, 115, 122 *et seq.*
Trapezoidal curve, 89
Tukey, J. W., 95, 134
Twisted J-shaped curve, 53, 94–5, 180
Type A expansion, 111–20
B expansion, 112–20

U-shaped curve, 41, 44, 52, 95, 119, 131
Uniform curve, 86, 88
Uven, M. J. van, 137

Variance, 9

Wold, H., 196